D0269857

Food Systems and Agrarian Change

Edited by Frederick H. Buttel, Billie R. DeWalt, and Per Pinstrup-Andersen

Searching for Rural Development: Labor Migration and Employment in Rural Mexico
by Merilee S. Grindle

Networking in International Agricultural Research
by Donald L. Plucknett, Nigel J. H. Smith, and Selcuk Ozgediz

Transforming Agriculture in Taiwan: The Experience of the Joint Commission on Rural Reconstruction
by Joseph A. Yager

NETWORKING IN INTERNATIONAL AGRICULTURAL RESEARCH

Donald L. Plucknett,
Nigel J. H. Smith,
and Selcuk Ozgediz

Cornell University Press

ITHACA AND LONDON

First published 1990 by Cornell University Press.

Printed in the United States of America

♾The paper used in this publication meets the minimum requirements of the American National Standard for Permanence of Paper for Printed Library Materials Z39.48–1984.

Library of Congress Cataloging-in-Publication Data

Plucknett, Donald L., 1931–
Networking in international agricultural research / Donald L. Plucknett, Nigel J. H. Smith, and Selcuk Ozgediz.
p. cm. — (Food systems and agrarian change)
Includes bibliographical references.
ISBN 0-8014-2384-8
1. Agriculture—Research—International cooperation. I. Smith, Nigel J. H., 1949– . II. Ozgediz, Selcuk, 1943– . III. Title. IV. Series.
S540.I56P58 1990
630'.72—dc20 90-31276

Contents

Foreword

Networking has emerged as a major means of furthering agricultural research around the world, particularly in developing countries. Indeed, the benefits of collaboration have spurred joint research efforts to improve and sustain crop and livestock yields in virtually every country. Given the heightened interest in networking to facilitate research, the publication of *Networking in International Agricultural Research* is timely.

Although few dispute that networks have improved the efficiency of agricultural research, there is some concern that too many networks could overburden the global agricultural research system, dampen enthusiasm for collaboration, and lead to more isolated efforts, thereby increasing redundancy. By painstakingly analyzing the ingredients for a successful network and offering solutions to problems in performance, *Networking in International Agricultural Research* provides a valuable service to the donor countries and institutions that fund international agricultural research programs, policy makers in agricultural research, and scientists.

Three very different disciplinary perspectives were harnessed to produce this book. Donald Plucknett is an agronomist, Nigel Smith a geographer, and Selcuk Ozgediz a management specialist. Donald Plucknett and Nigel Smith constructed some of the conceptual and historical framework for the book in an article they wrote on networking for *Science* in 1984. At the same time, Selcuk Ozgediz was working on the management of agricultural research institutions, in particular the international centers supported by the Consultative Group on

International Agricultural Research (CGIAR). Thus, their book reflects the convergence of separate strands of inquiry into the study of networks.

The book is strongly conceptual as well as descriptive. The history of networking is reviewed, followed by a typology of networks. Discussion of the organization, management, and development stages of networks is likely to be particularly helpful to administrators of agricultural research. Chapters on principles for effective networking and problems to be avoided are beacons to steer networking efforts toward more productive courses. Chapters on the four major types of networks will help clarify the aims and scope of each network category. Networks as international agricultural research centers (NIARCs) is a term coined in this book to describe an organization genre that surfaced on the agricultural research scene during the 1980s. The discussion of NIARCs and the final chapter on the future of networking are full of insights and are especially relevant to policy makers.

Networking in International Agricultural Research is relevant to readers outside of the agricultural research community as well. The rapid advances in information technology since the 1960s will likely change the way we do business in the future. Traditional forms of organizations will give way to new forms. The place where we work will become less and less important as we are better able to communicate effectively with one another. Our recent experience with networking as an organizational form, therefore, has much light to throw on how we can conduct business in the future.

W. David Hopper

The World Bank
Washington, D.C.

Preface

Networking has become popular in many human activities, ranging from group advocacy of public interest causes to research and development efforts in industrial and scientific establishments. People now talk of "networking" with others who have similar interests and goals. Social networks provide companionship and support; business colleagues use networks to further their careers. Scientists are increasingly involved in networks to exchange information, discuss problems, and plan research.

Networking is a new name for an ancient practice. People have cooperated with one another since the beginning of the human race. But the extent and organizational modes of collaboration have changed markedly since the 1960s particularly in science. Research networks are proliferating, aided in part by new technologies that facilitate communication among scientists.

Nowhere is networking better developed than in agricultural research, the focus of our study. Networks now operate in virtually all areas of agricultural research, at both the regional and international levels. Networking has gained momentum because it promises increased efficiency in research, a valuable asset in an age of tightening resources. By dividing up the task and sharing information on results, networks can make research more efficient. An important characteristic of successful networks is that solutions to widespread problems are usually found earlier than if individual scientists or institutions work separately. And if the problem is intractable with present knowledge and technology, at least that conclusion is reached more quickly and cheaply by networking than when research is conducted in an isolated, piecemeal fashion.

Given the global proliferation of networks, a review of their history, performance, principles for success, problems and remedial measures, and future directions should be helpful, particularly to policy makers. The major goals of our survey of agricultural networks are to elucidate the ingredients for successful collaboration, to uncover problems, and to suggest corrective actions. Although a cooperative spirit is an asset in agricultural research, goodwill alone will not guarantee productive and high-quality work. We therefore adopt a critical, evenhanded look at networks, rather than an advocacy approach.

In discussing networks, we present examples from the experience of international agricultural research centers and national institutions around the world. Our aim is to provide policy makers in research, particularly in the agricultural sciences, with information that will help them decide to what extent networks would facilitate their programs. Scientists should also benefit from our examination of the payoffs and pitfalls inherent in collaborative research. Networking takes time, and the benefits must outweigh the costs if research momentum is to be maintained.

In Chapter 1, "Rationale and Roots," we profile the major themes and questions we intend to address throughout the book. We outline the different ways the concept of networking is used and emphasize our concentration on collaborative research. We trace the origins of networking in agricultural research, from European colonial times to the advent in the 1960s of international agricultural research centers working in collaboration with national programs.

The various categories of networks are outlined in Chapter 2, "Typology," along with an explanation of the typology adopted for our review of agricultural research networks. In Chapter 3, "A Conceptual Framework for Studying Network Effectiveness," we offer a conceptual model for examining factors contributing to the effectiveness of networks. The conceptual model guides our discussion of governance and management mechanisms employed by various networks and organizational structures adopted to enhance research efforts.

In Chapter 4, "Development Stages," we explore the origins of networks and trace their trajectory from start-up to maturity. We then bring together the essential ingredients for viable research networks in Chapter 5, "Principles for Success," in which we explore such issues as funding, leadership, and motivation. The importance of clearly defining the research problem as well as using sound methodology is stressed. Our discussion of principles is illustrated with concrete examples and provides a framework for assessing the payoffs and problems in networking.

Networking has penetrated virtually every facet of agricultural research, and in Chapters 6 through 9 we examine the performance and problems of networks concerned with a wide array of agricultural research fields. Our description and analysis of the results and difficulties of networking follow the typology laid out in Chapter 2. Thus in Chapter 6, "Information Exchange Networks," we sample networks set up to disseminate information. Chapter 7, "Material Exchange Networks," focuses on networks established primarily to exchange and test plant germplasm and agricultural machinery. Chapter 8, "Scientific Consultation Networks," covers independently planned research programs that agree to cooperate in some aspects of their work. In Chapter 9, "Collaborative Research Networks," we discuss research efforts that are jointly planned and that adopt uniform research methodologies. The performance, management, and major contributions of each of the network categories are discussed.

Some international agricultural research centers have only limited facilities at headquarters and therefore operate mostly by networking. But because some international research problems demand a more decentralized approach, we describe in Chapter 10, "Networks as International Centers," the operations of several recent organizations that combine the advantages of both research institutions and networks. Networks as international agricultural research centers (NIARCs) mobilize their staffs and resources in association with national programs to tackle a variety of topics, ranging from improving varieties of bananas and plantains to soil and water management.

Problems that have arisen in networking are discussed in Chapter 11, "Problems and Remedies." We not only identify lapses in the performance of networks but show how they are, or can be, overcome through flexible and innovative management. In the broadest sense, problems in networking occur when participants depart from principles for success. Indeed, the scientific, personnel, and institutional difficulties we pinpoint in our examination of the different types of networks reinforce the principles for success outlined earlier in the book. This policy-related chapter tests the validity of the principles for success, exposes pitfalls to be avoided, and points out areas that scientists, administrators, and donors must focus on to improve existing networks.

In the final chapter, "Lessons and Future Directions," we explore such issues as whether the global research system is overloaded with networks and how communication between networks can be improved. We address the larger question of whether networks can adequately substitute for existing or new international agricultural research centers. Networks on the drawing board are discussed to illustrate the ever-changing panorama of networking and to highlight some perceived needs in agricultural research.

Rapid changes in computer technologies are constantly creating new opportunities for networking, and we indicate the various ways in which computers facilitate collaborative research. We also explore various computer hardware and software configurations used by international and national agricultural research institutions to facilitate networking.

The importance of training for Third World scientists and technicians is stressed because a network is only as good as its participants. And we discuss a topic related to training: the dynamic nature of the networking relationships between international research centers and national programs.

Scientists and other decision makers currently involved in networks, and those thinking of joining or starting a network, will find policy-related concepts and conclusions throughout the book. Networking is no panacea for lagging agricultural production or inefficient agricultural research, but it can be a powerful way of improving the quality and impact of research.

Many individuals have kindly shared their ideas and thoughts about networking in agricultural research with us. To the extent possible, we cite specific papers and published accounts to credit sources of information and insights on networking. In 1983, Peter Greening suggested that networking would be a useful research topic to pursue. We also thank the following individuals for exchanging their observations on agricultural research networks with us: Robert Bertram, Nyle Brady, Ken Brown, Virgilio Carangal, Michael Collinson, Ralph Cummings, Jr., Dana Dalrymple, William Furtick, Dennis Greenland, Robert Herdt, Guy d'Ieteren, Calvin Martin, Gustavo Nores, Alexander von der Osten, D. V. Seshu, John Trail, and Donald Winkelmann. Three anonymous readers selected by Cornell University Press made helpful suggestions on an early version of the manuscript. Finally, we are grateful to the United Nations Development Program (UNDP) and the United States Agency for International Development (USAID) for supporting research on the book.

The views and opinions expressed in the book are ours and do not necessarily indicate the approval of any reviewers or organizations.

D. P., N. S., AND S. O.

Washington, D.C.
Gainesville, Florida

Networking in International Agricultural Research

1

Rationale and Roots

"Networking" has been firmly established in the English language since at least the 1960s. At first confined to technical matters, such as telephone networks, the concept of networking has spilled over into social affairs, business, and research. A network can be defined as an association of independent individuals or institutions with a shared purpose or goal, whose members contribute resources and participate in two-way exchanges or communication. Unlike bureaucracies, networks usually do not have a well-defined hierarchy of authority, although they do have levels (Foley, 1989). An important feature of a network is its decentralized nature.

The networking concept is simple and relatively easy to apply, hence its power and ready acceptance. Individuals and institutions are increasingly realizing that they operate in a world in which nations depend on each other more and more for economic, cultural, and scientific benefits.

Improvements in crop yields and farm management practices have always depended on a generous international flow of planting materials, implements, and ideas. Agricultural research is increasingly a global undertaking, and the continued productivity of farmlands on every continent and island rests to a large extent on joint efforts to uphold and improve yields. Sustainable agriculture depends on inputs from a broad range of disciplines, so collaborative research teams that tap scientists from a wide variety of fields have come together to help boost crop and livestock production and to minimize yield fluctuations (Figure 1). Isolated research programs are finding it increasingly dif-

Figure 1. Scientists from the International Trypanotolerance Center talking with Fulani farmers in Gambia who are participating in trials that assess cattle response to tsetse fly challenge. The scientists are collaborators in the Trypanotolerant Livestock Network.

ficult to confront alone the numerous challenges to boosting food production.

Although we focus on agricultural research networks, international collaboration is common in other scientific fields, particularly astronomy, astrophysics, and medical research. For example, a massive supernova explosion in the Large Magellanic Cloud in February 1987 was observed by an international network of stations. Japanese and U.S. scientists have been working particularly closely together in observing this supernova, and so have the Russians and Italians (Garwin, 1987). In research on polio, collaboration between scientists in the United States, Czechoslovakia, and the Soviet Union was instrumental in screening millions of tissue cultures of live polio viruses (Sabin, 1986). Such active research cooperation led to the release of live polio virus vaccines in 1960, much earlier than if scientists had worked in isolation. Researchers at Texas A&M University and the University of Idaho are setting up a collaborative research network to investigate the potential of baculovirus, an insect virus, as a vector for developing vaccines for hepatitis B, acquired immune deficiency syndrome (AIDS), and malaria (Wright, 1986). And recent outbreaks of a puzzling disease that resembles mononucleosis in the Lake Tahoe area of California and Nevada have spurred collaboration among investigators in laboratories across the United States (Kolata, 1986).

Scientists in Western Europe are also working together in a wide array of research endeavors. The European Science Foundation alone has fostered nine research networks covering such topics as polar science and longitudinal studies of individual development (Dickson, 1987; Maddox, 1987). Major collaborative research projects in Europe include the European Laboratory for Particle Physics (CERN), which has fourteen member states; the Brussels-based Basic Research in Industrial Technologies in Europe (BRITE), established by the European Economic Commission in 1985 to support precompetitive industrial research; the European Molecular Biology Laboratory (EMBL), started in 1974 in Heidelberg, West Germany, with fourteen member states; the European Southern Observatory (ESO), set up in 1962 and headquartered in Munich, with eight participating countries; and the Joint European Torus (JET), established in 1977 at Culham, England, which unites fifty-eight research institutions in twelve nations in the study of plasma for the production of fusion energy.[1] The European Science Foundation was also instrumental in launching the European Network for Research on the Neural Mechanisms of Learning and Memory in 1988.[2]

Although international cooperation is increasingly common in many scientific fields, networking is probably best developed in agricultural research. Government leaders, administrators, and scientists concerned with alleviating worldwide hunger and improving the livelihoods for rural peoples increasingly realize that they often need to turn to other nations for help. Assistance is frequently required in the form of resistant plant material to combat a disease outbreak or new machine designs that are both energy efficient and affordable. Agricultural research is thus a particularly fertile area to examine networking and explore issues and problems associated with collaborative research.

Crop breeders, for example, depend on the importation of fresh plant material for improved traits. The pace of germplasm exchange is accelerating as plant breeders seek fresh options for developing new

1. *Science* 237 (1987):1113.
2. *Nature* 330 (1988):793.

Figure 2. Collaborators in the International Network on Genetic Enhancement of Rice, formerly known as International Rice Testing Program, reviewing deepwater rice trials at Huntra Station, Thailand. Courtesy of the International Rice Research Institute.

crop varieties to sustain yields in modern farming areas and to tailor varieties to difficult environments (Figure 2). International nurseries, one of the most widespread forms of networks, in which breeders assess the performance of plant material in a wide range of environments, have emerged as major launching pads for new crop varieties around the world. International nurseries are now indispensable to the global effort to sustain agricultural productivity.

Both international and regional networks have sprung up to further agricultural research in a variety of other areas. Network teams are currently investigating such topics as the transferability of technology between similar soils in different regions, the response of crops to various fertilizer treatments, the efficacy of biological nitrogen-fixing systems, the suitability of various farm machinery designs, biotechnology, and farming systems. In livestock research, networks are investigating such subjects as the use of agricultural by-products for cattle feed in the tropics and the epidemiology of trypanosomiasis, a debilitating livestock disease present in large areas of Africa.

More than a hundred international agricultural research networks are currently operating (Appendix 1), and even though some fold after a few years, there has been a marked net increase in their number. When one considers subnetworks under the umbrella of larger collaborative efforts, particularly nurseries, the number of international agricultural research networks may easily exceed two hundred or more.

Networking has spread to many frontiers of agricultural research because of the perceived benefits of collaborative research. These dividends include greater efficiency of research, cross-fertilization of ideas, the ability to use existing facilities rather than building new ones, and the pooling of research talent rather than adding staff. Another powerful motivating force is the opportunity to attend international workshops, accompany monitoring tours, and obtain further training.

Although this trend toward international collaborative research has many advantages, the proliferation of networks all over the world is consuming the time and resources of thousands of scientists and hundreds of institutions. An assessment of the tangible products and the improvement of human resources resulting from networking in agricultural research is warranted.

Not everyone is convinced of the merits of more networking in international agricultural research, or even of the value of many existing networks. Skeptics raise some basic questions. Has networking extended so far that it is now actually entangling and constraining research? Have some collaborative research efforts accumulated too many side projects, making them unwieldy? Are some networks designed to meet the needs of sponsoring organizations or coordinators rather than participants? Other central issues we explore include the extent to which networks can effectively substitute for the creation of new research centers, whether some networks may have outlived their usefulness, and whether collaboration rather than competition between research programs is more effective in delivering high-quality products and services.

As with many other good ideas, there is always the risk of a bandwagon effect in the creation of agricultural research networks. Some networks may be more "net" than "work." Just as committees are sometimes appointed to show that action is being taken to solve an intractable problem, networks sometimes are hastily assembled without carefully defining the problem and drawing up a realistic work schedule. Also, just as committees can become an excuse for inaction, networks can become ineffective or paralyzed unless they are founded on sound principles. A close examination of the bedrock principles for successful networking will help guide policy makers when establishing or fine-tuning networks.

History

During the eighteenth and nineteenth centuries, the Belgians, British, Dutch, and French established networks of agricultural research stations in their tropical and subtropical territories, mostly to increase the flow of export crops. Botanic gardens established by colonial powers played prominent roles in plant introduction and multiplication. The Royal Botanic Gardens, Kew, for example, served as a hub in a network of British botanic gardens spanning the Caribbean and Asia (Smith, 1986). British and Dutch botanic gardens in the tropics introduced rubber (*Hevea brasiliensis*),[3] cinchona (*Cinchona* spp.) and African oil palm (*Elaeis guineensis*) to Southeast Asia and promoted the establishment of commercial plantations in Malaysia, Indonesia, and Sri Lanka. Colonial stations improved the output of cotton (*Gossypium* spp.), groundnut (*Arachis hypogaea*), and sugarcane (*Saccharum* spp.), but coordination and collaboration with territories outside the individual empires were limited (Desai, 1982). Furthermore, many stations closed after independence because they were staffed mainly by expatriates (Eicher, 1982).

In the United States, networking in agricultural research had its roots in informal groups working together on a regional basis. The continental expanse of the United States and its varied farming conditions created a need to link research at the state, regional, and national levels. To increase agricultural productivity across the ecologically diverse country, an integrated, two-tier system arose: the United States Department of Agriculture (USDA) at the federal level working in conjunction with state-run agricultural experiment stations. The two systems synchronized research programs on widely shared problems. For example, researchers working on hard red spring wheat at state experiment stations in the northern Great Plains and the North Central region coordinated their efforts during the 1920s (Moseman, 1970:48). Breeders, plant pathologists, biochemists, soil scientists, and entomologists also joined forces to further research on hard red winter wheat in the Great Plains, soft red winter wheat in the East, and white wheat in the Pacific Northwest.

Networking provided a springboard for the development of hybrid maize (*Zea mays*) in the United States. In 1925, researchers working at state agricultural experiment stations in the Cornbelt coordinated

3. The first time a plant or animal is mentioned its scientific name is usually given; thereafter only the common name is used.

their programs to avoid duplication and to share results promptly. This network, coordinated by the USDA under the authority and inspiration of the Purnell Act, led to the release of commercial hybrids in the 1930s which rapidly transformed maize farming in the United States (Hayes, 1963; Moseman, 1970). Because hybrid maize generally outyields traditional, open-pollinated varieties, hybrids would have been developed without this network. But hybrid maize seed reached farmers' hands more quickly as a result of networking.

Before World War II scientists exchanged a limited amount of plant germplasm across national boundaries, but for the most part agricultural networks were informal and national in scope. A serious outbreak of a new race (15B) of wheat stem rust in the United States during the early 1950s triggered the establishment of the first formal, multinational network to screen crop germplasm. The International Stem Rust Nursery, initiated in 1950 and coordinated by the USDA, was the first systematic nursery to transcend national borders (Loegering and Borlaug, 1966; CIMMYT, 1979). This pioneer nursery spearheaded wide efforts to share and evaluate wheat materials. Wheat breeders were eager to participate because they realized that a neighbor's problem today could be theirs tomorrow. Initially, researchers in the United States, Canada, Mexico, Colombia, Ecuador, Peru, Chile, and Argentina benefited by sharing and evaluating wheat materials. By 1970, 150 scientists in 40 countries in the Americas, Europe, and the Middle East had joined the network (Moseman, 1970).

Another early network, Programa Cooperativo Centroamericano para el Mejoramiento de Cultivos Alimenticios (PCCMCA), grew out of a meeting in Costa Rica in 1954 (R. Havener, pers. comm.). Originally, the Central American network of scientists and administrators focused on improving maize production, but by 1964 beans, sorghum (*Sorghum bicolor*), rice (*Oryza sativa*), root and horticultural crops, as well as livestock, were included. During its embryonic stage, only twenty to thirty scientists participated in the network, but PCCMCA now has some five hundred members. This network has expanded into the Caribbean, and representatives from several South American countries attend its annual meeting. To facilitate the exchange of crop germplasm and other technologies, PCCMCA has forged working relationships with several international agricultural research centers (Map 1) such as the International Maize and Wheat Improvement Center (CIMMYT—Centro Internacional de Mejoramiento de Maiz y Trigo), the International Center for Tropical Agriculture (CIAT—

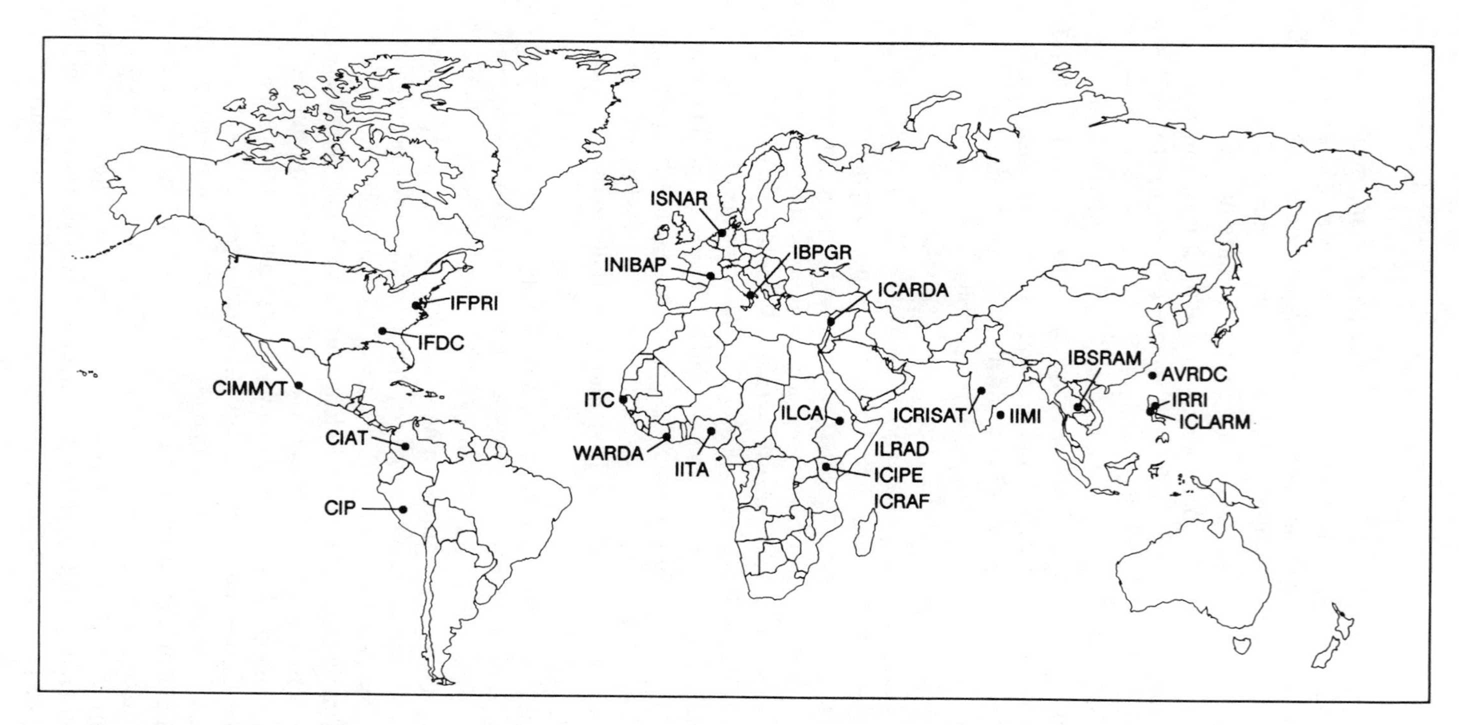

Map 1. International agricultural research centers that are involved in international agricultural research networks. See Appendix 3 for acronyms, founding dates, and research programs of the centers.

Centro Internacional de Agricultura Tropical), the International Potato Center (CIP—Centro Internacional de la Papa), and the International Crops Research Institute for the Semi-Arid Tropics (ICRISAT).

With the assistance of the Rockefeller Foundation, India began its All-India coordinated crop improvement programs in the late 1950s. The first of these systematic national networks were devoted to maize, sorghum, and millets; in the 1960s the concept was extended to include rice, wheat, and other crops. The Rockefeller Foundation provided the joint coordinator for the program, in which researchers worked on a specific crop in the Indian Council for Agricultural Research (ICAR) and individual states within India. Researchers reviewed results of trials at annual workshops, and coordinators made periodic visits to network sites. Each network published an annual report. Crop breeders were thus able to judge at a glance the best performers across a wide range of conditions and spot outstanding lines that matched their needs. The following year's programs were planned and specific responsibilities added.

The Food and Agriculture Organization (FAO) of the United Nations and the U.N. Development Program (UNDP) have employed a network approach to promote agricultural development since they started soon after World War II. FAO has encouraged networks to foster research and technical collaboration in various fields, to upgrade national research capabilities, and to arrange for training and information exchange. UNDP has supported many of these activities. The FAO/UNDP Regional Project on Field Food Crops in Near East and North African countries has collaborated with CIMMYT and other institutions since 1963 in testing improved cereal lines and in training scientists. FAO has encouraged the development of ten agricultural research networks in Europe dealing with such commodities as maize, sunflower (*Helianthus annuus*), olive (*Olea europaea*), soybean (*Glycine max*), sheep, and pastures since 1975. A similar approach is followed in Latin America. In Asia research and development networks concentrate on water buffalo (*Bubalus bubalis*), coconut (*Cocos nucifera*), root crops, organic fertilizers, and farm machinery. In Africa, FAO assists agroindustrial networks and collaborative research on trypanosomiasis.

United Nations agencies actively support other agricultural networks. Following the 1972 Stockholm Conference on Man and the Environment, the U.N. Educational, Scientific and Cultural Organization (UNESCO) and the U.N. Environment Program (UNEP) organized a series of Micro-Biological Resources Centers (MIRCENs).

MIRCENs link research institutes in regional cells in East Africa, the Middle East, Southeast Asia, and Latin America and work on biological nitrogen fixation, waste recycling, biological control of pests, and industrial fermentation. In 1976, MIRCENs began systematically organizing gene banks for microorganisms important in agriculture, fuel production, and the food and brewing industries.

The Canadian and U.S. governments also actively support networking in international agricultural research. Canada's International Development Research Centre (IDRC) has demonstrated a strong commitment to networking in agricultural sciences and other research areas spanning several decades. The Board for International Food and Agricultural Development (BIFAD) and the U.S. Agency for International Development (USAID) in Washington, D.C., also participate in networking in the Third World. The Collaborative Research Support Programs (CRSPs), assisted by BIFAD in Washington, D.C., with funding provided mainly by USAID, encompass a series of multipurpose commodity networks linking U.S. institutions with counterparts in developing countries. Their goal is to match the agricultural research interests of U.S. universities with similar research thrusts in developing countries (Wennergren et al., 1986). Self-interest and humanitarian concerns are the two major motives for participation in the various CRSP networks, which began with the small ruminants program in 1978.

Impetus for the establishment of CRSPs came from Title XII of the Foreign Assistance Act, also known as the Famine Prevention and Freedom from Hunger Act. This far-reaching legislation was designed to extend to impoverished nations the land-grant education, research, and extension system that has worked so well for U.S. agriculture in the last one hundred years.

At present forty U.S. universities are working with sixty-six agricultural institutions in thirty developing countries in the various CRSP programs. Some nine hundred scientists are engaged in CRSP activities worldwide. On the basis of recommendations of a world food study conducted by the U.S. National Academy of Sciences in 1976, eight mutually beneficial CRSPs were set up to work on small ruminants (beginning in 1978), sorghum and millet (1979), bean and cowpea (1980), soil management (1981), nutrition (1981), groundnut (1982), pond dynamics (1982), and fisheries stock assessment (1985). Each CRSP has a management entity, a technical committee, a board, and an external evaluation panel to provide guidance and quality control.

International agricultural research centers, many of them under the umbrella of the Consultative Group on International Agricultural

Research (CGIAR), have acted as catalysts for numerous agricultural research networks in the Third World since the 1960s (Plucknett and Smith, 1982; CGIAR, 1985). International agricultural research centers have been especially active in setting up international nurseries to evaluate the genetic potential of cereals, pulses, root crops, and forage plants on every continent. International nurseries are a relatively recent development, since most international agricultural research centers were established in the 1960s and 1970s (Appendix 3). The Philippines-based International Rice Research Institute (IRRI), for example, began operating in 1960 and initiated the first international rice nursery in 1963. In 1964, the Rockefeller Foundation–financed forerunner to CIMMYT organized the International Spring Wheat Yield Nursery (ISWYN) by merging two regional nurseries, the Inter-American Nursery Trials, which started in 1960, and the Near East–American Spring Wheat Yield Nursery, which began in 1962. The newly formed ISWYN spanned thirty-four sites in twenty-three countries in the Americas, Africa, and Southwest Asia (Moseman, 1970:81).

Three international agricultural research centers grew out of networks or were established specifically to promote networking. The International Center for Agricultural Research in the Dry Areas (ICARDA), based in Aleppo, Syria, picked up the network strands laid out from 1968 to 1976 by the Arid Lands Agricultural Development Program (ALAD). ALAD was established with the help of a consortium of donors led by the Ford Foundation to boost agricultural yields in the Middle East. Like most networks, ALAD was a mechanism rather than an institution; its main purpose was to test germplasm of cereals and pulses and to boost sheep production. National programs conducted the research at widely scattered sites. ALAD thus had laid much of the groundwork for regional nurseries when ICARDA was established in 1976.

The idea of network coordination was also central to the mission of the West Africa Rice Development Association (WARDA) and the International Board for Plant Genetic Resources (IBPGR). WARDA, headquartered near Bouake, Côte d'Ivoire, assists in linking rice researchers in fifteen countries of West Africa and helps run international nurseries in member countries. Rome-based IBPGR serves as a catalyst for the collection and conservation of crop germplasm and acts as a planning and information nerve center. IBPGR also supports studies on the distribution of crop genetic resources to help establish priorities for collecting.

With a mandate to foster the establishment of gene banks and to help coordinate the collection and storage of germplasm, IBPGR has established minimum standards for adequate gene banking and has forged a network of base collections for certain crops. Base collections are gene banks that contain a broad representative sample of the genetic diversity of a crop.

The number of base collections in IBPGR's global network has risen steeply. In 1976, only five base collections were in the network, but by 1983 the number had risen to thirty. As of 1985, thirty-three base collections in twenty-four countries were in the network representing thirty-four staple seed crops (Plucknett et al., 1987). IBPGR has set a goal of fifty base collections for seed crops. When base collections in field gene banks are included, the number of base collections in IBPGR's network climbs to forty-six, twenty-nine of which are in the Third World. IBPGR is also considering a formal network of medium-term collections, which, following its pragmatic approach with long-term gene banks, would involve better coordination and some upgrading of existing collections.

Gene banks are linchpins in the global effort to improve and stabilize crop yields, and more than seventy countries now have medium- or long-term crop germplasm collections. Gene banks make a considerable, albeit indirect, contribution to international nurseries because breeders constantly tap germplasm collections to enrich their breeding pools. IBPGR's role in furthering gene bank networks thus underpins a much more extensive network of international nurseries to the benefit of farmers and consumers worldwide.

2

Typology

In its broadest sense, networking implies the linking of individuals or institutions with a shared purpose to achieve desired goals. A diverse array of networks, sometimes with extensive and intricate linkages, have sprung up to address a variety of needs and purposes. Some networks are informal arrangements in which a handful of participants remain in contact mainly through newsletters, often published by desktop computers. Other, more elaborate research networks are linked by electronic mail, abide by commonly agreed-upon research protocols, involve regular international travel, and train scores of individuals.

Most researchers have colleagues with whom they correspond and exchange ideas and information, and some may call this arrangement a network. These usually involve a loose web of intermittent contacts with no clearly defined hub or nerve center, a normal feature of networks. These informal communications can evolve into more formal structures. Informal networks tend to begin simply and gather momentum and complexity as participants learn from each other. Indeed, the systematic sharing of information and technologies is one of the greatest strengths of networking.

In networks, scientists can participate where they are, as they are, with benefits to all. In time, a network may come to resemble an intensive team research effort, but most do not start that way. In embryonic networks, communication is likely to be one-way rather than bidirectional, and commitment of resources, both from outside sources and participants, may be limited.

Although all forms of networking can be valuable, we concentrate here on active, rather than passive, networks. By active we mean that participants commit some of their resources, exchange information and technologies with each other, and have some inputs in setting priorities and planning. Active participants do more than simply receive information, machines, or plant germplasm. They try out ideas and technologies and report back to colleagues or a coordinator so that other participants can benefit from feedback.

Before exploring conceptual models for networks, we examine various attempts to classify networks. We have adopted a typology that builds on earlier efforts to categorize networks. Our typology provides a framework for discussing conceptual models for networks and for exploring principles, payoffs, and problems of networks which are discussed in later chapters.

Network typologies can be established according to purpose, major product or service, or operational style. Ideally, each category should be mutually exclusive so that a given network fits into only one category. In practice, however, some networks are difficult to categorize because they may be evolving into new forms and may well fit into more than one mold. A balance must always be struck between a typology that is too simplistic and one that is so fine-grained that the number of categories is virtually inexhaustible.

Any number of typologies can be drawn up to classify agricultural research networks, depending on the purpose and analytical goals. For example, Plucknett and Smith (1984) grouped agricultural research networks by subject matter, such as international nurseries and farming systems networks. One advantage of this typology is that individuals interested in particular fields of research can quickly locate the appropriate parts of the text relevant to their interests.

Ralph W. Cummings, Jr., and Calvin Martin of the U.S. Agency for International Development categorized agricultural research networks according to their operational style in a report to the Special Program for African Agricultural Research (SPAAR, 1986). Cummings and Martin propose three main types of networks, commonly referred to as the SPAAR typology: information networks (type 1), scientific consultation networks (type 2), and collaborative research networks (type 3). This typology is very useful and has been widely understood and accepted.

We modify the SPAAR typology slightly by adding a further classification: material exchange networks. The typology employed for our descriptive and analytical purposes therefore comprises the following categories: information exchange networks, material exchange networks, scientific consultation networks, and collaborative research networks.

Information Exchange Networks

In their simplest form, information exchange networks are essentially passive operations in which information is fed out from a coordinator to individuals on a mailing list. Information in such networks is disseminated largely by newsletter. A two-way flow occasionally develops when individuals on the mailing list provide items to the coordinator to pass on to others in the network. In most cases, though, members merely have their name put on a mailing list, usually without cost, and rarely interact with others in the network. Contact with the coordinator is typically limited to periodic updating of the mailing list.

Material Exchange Networks

Material exchange networks are primarily concerned with testing crop protovarieties and agricultural machinery designs at various sites. International nurseries for testing crop germplasm and agricultural machinery networks fall under this category. Participants in international nurseries plant mostly the same materials and use identical screening methodologies. In this manner, superior germplasm with wide adaptability and resistance to diseases and pests can be identified more reliably.

We do not wish to imply that material exchange networks are mundane operations, devoid of research or intellectual input. Materials are exchanged to bolster research efforts. We establish this category merely to point out that the major focus of some networks is the exchange and testing of plant materials or machinery prototypes. Material exchange networks serve as wholesalers and evaluators for other research programs.

Scientific Consultation Networks

In scientific consultation networks, researchers involved in preexisting, autonomous projects agree to share information and ideas. Scientists working in isolation increasingly realize that new insights can be gained by sharing experiences. Changes may be implemented in independent research programs as a result of interaction with colleagues in the network. Workshops or conferences are often organized on a regular basis to facilitate discussion of research progress and problems (Table 1). Research projects are still locally planned, however, and participants may use different methodologies. Scientific consultation networks are cooperative rather than truly collaborative undertakings.

Scientific consultation networks are not as difficult to set up as collaborative research networks because research programs are already in place and are linked in the network by exchanging information and ideas. But the ultimate goal of many scientific consultation networks is collaborative research. Such a step is not easy. Scientists are busy with their current research activities and may be reluctant to restructure their research approach radically to accommodate jointly planned ventures. Because of the inertia and disparate nature of ongoing research projects, it is therefore sometimes easier to assemble a collaborative research program from scratch.

Collaborative Research Networks

The major distinguishing feature of collaborative research networks is that research is jointly planned and carried out. Most collaborative research networks are new ventures, at least in scale and scope, even though they use existing facilities and personnel, and are planned by all members. Research methodology is generally more uniform than in

Table 1. Some characteristics of networks that illustrate the operations and levels of complexity of information exchange, material exchange, scientific consultation, and collaborative research

Trait	Information exchange	Material exchange	Scientific consultation	Collaborative research
Coordinator	Yes	Yes	Yes	Yes
Publications	Yes	Yes	Yes	Yes
Advisory board	No	Yes	Yes	Yes
Study tours	No	Yes	Yes	Yes
Training	No	Yes	Yes	Yes
Workshops	No	No	Yes	Yes
Common methodology	No	Yes	No	Yes
Joint planning	No	No	No	Yes

the case of scientific consultation networks. Participants in collaborative research networks are actively involved in establishing research priorities, dividing responsibilities, and following common methodologies.

Although all types of networks have important roles to play in furthering research and improving agricultural productivity, collaborative research networks have the greatest potential for upgrading skills of participants and helping Third World nations overcome constraints to increased agricultural production. Scientists in collaborative research networks adopt the same, or at least very similar, methodology, and results may have broad appplication and are generally transferable.

In a sense, collaborative research networks represent networking in its purest form because individuals in jointly planned research ventures pool their resources and talents, participate in planning and policy formulation, and work in unison. But not all networks can, or should, become collaborative research efforts. For some purposes, information exchange, material exchange, or scientific consultation networks suffice. But for other research problems, particularly if they are narrowly focused, collaboration can be a powerful approach. Less than a dozen genuine collaborative research networks were in operation in 1989.

The Trypanotolerant Livestock Network in Africa exemplifies collaborative research and has many lessons to offer. Other jointly planned research networks whose track records can guide the establishment of future collaborative research programs include the International Benchmark Sites Network for Agrotechnology Transfer (IBSNAT), the Asian Rice Farming Systems Network (ARFSN), and Programa Regional Cooperative de Papa (PRECODEPA), a multipurpose commodity network focusing on potato production in Central America and the Caribbean.

The Trypanotolerant Livestock Network, coordinated and supported by the International Livestock Center for Africa (ILCA) and the International Laboratory for Research on Animal Diseases (ILRAD), focuses on a widespread African livestock disease, trypanosomiasis. IBSNAT, coordinated and supported by USAID and the University of Hawaii, has built on the work of a predecessor research project, the Benchmark Soils Project. IBSNAT, assembled in 1982, explores the transferability of agronomic practices and cropping systems within three major soil families of the tropics. The International Rice Research Institute helps administer ARFSN, a farming systems network

Figure 3. Participants in the PRECODEPA regional potato network interviewing a farmer near Toluca, Mexico. Courtesy of the International Potato Center.

that has grown spectacularly in its dozen years of existence and now spans fifteen Asian nations. ARFSN embraces several research thrusts, including integration of livestock in rice farming systems, fertilizer trials, and varietal testing. PRECODEPA, one of five regional potato research networks organized by the Lima-based International Potato Center, addresses ecological and socioeconomic constraints to increased potato production in nine Central American and Caribbean nations (Figure 3). PRECODEPA has carved up the research responsibilities as follows: late blight (Mexico); seed production (Mexico, Guatemala, Costa Rica); tuber moth (Costa Rica, Guatemala); golden nematode (Panama, Mexico); bacterial diseases (Costa Rica); rustic storage (Guatemala); socioeconomics (Guatemala); and processing (Guatemala).

The typing of networks can be a useful tool in their planning and management. Common characteristics and problems can be identified, and a conceptual structure can be identified to help plan and evaluate network effectiveness.

3

A Conceptual Framework for Studying Network Effectiveness

A common misconception is that networks are not organizations. Organizations are usually defined as a collection of individuals or units working toward a common purpose or goal (Etzioni, 1964). Networks are an organizational form, albeit less bureaucratic and hierarchical than institutions such as universities or corporations.

The literature on management of organizations offers many models that attempt to explain organizational performance. Models proposed by Pascale and Athos (1981), Paul (1982), Tichy (1983), Killman and Saxton (1984), and Egan (1988) are notable studies of factors contributing to effective management of organizations.

The conceptual framework we propose for studying network effectiveness builds on elements of previous models and has been honed from studying the effectiveness of international agricultural research institutions (Ozgediz, 1987, 1990). Our conceptual model has five major components: network guidance, management of network resources, management of network activities, management skills and teamwork, and a network's linkages with its environment (Chart 1).

Network Guidance

Guidance involves establishing the purpose and direction of the network's activities as well as setting broad policies to help steer the partners' joint efforts. Network guidance incorporates four factors—guiding values, leadership, governance, and strategy.

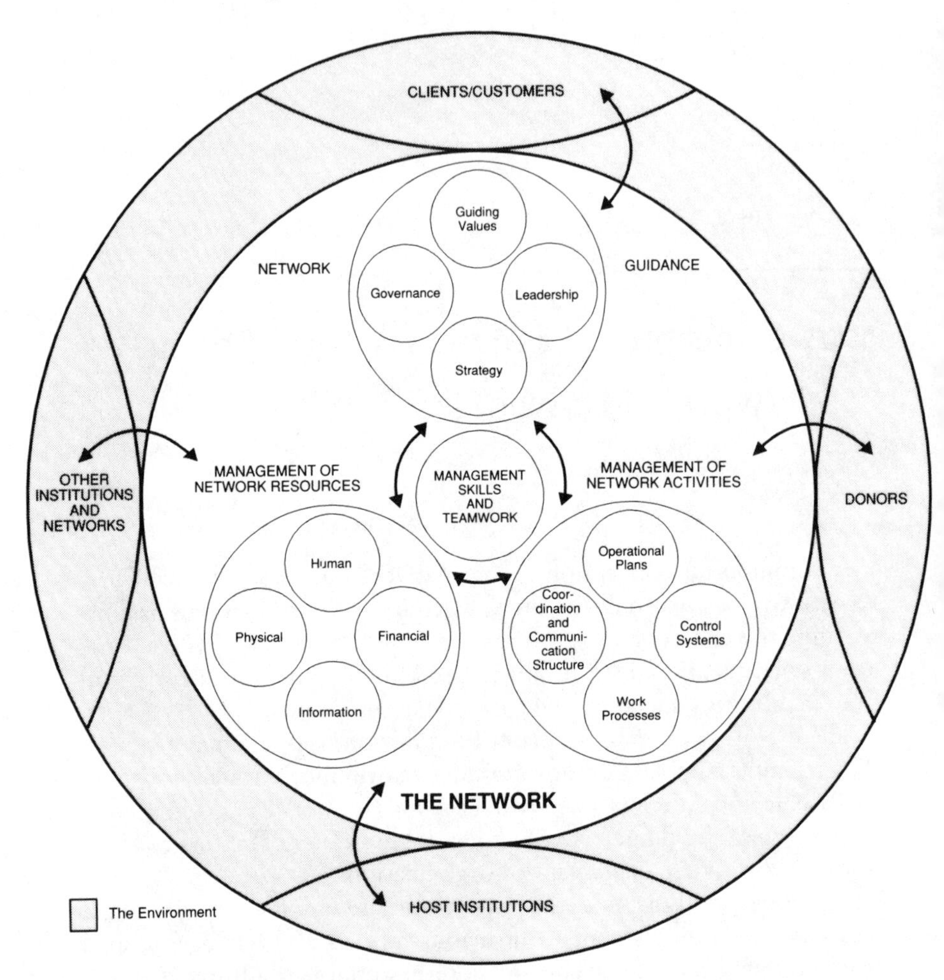

Chart 1. Conceptual model for studying the effectiveness of a network

Guiding Values

Guiding values are principles and basic ideas shared by members of the network. Values such as the degree to which the network is oriented to clients, the degree of autonomy of network members, attention to quality in research, team spirit, encouragement of innovation, and the impact of political discord are important in shaping the actions of network members. Some guiding values are set by the leaders of the network, while others emerge as the network operates.

Although the interacting nodes or participants in a network will have their own subcultures, the norms of behavior and operational style are heavily influenced by the coordinating body. Many agricultural research networks are coordinated by international centers, each of which has its own culture. Some of the values and operational style of the coordinating center inevitably suffuse into the network.

The International Potato Center, for example, has always emphasized a decentralized, regional approach. CIP operates from its modest headquarters in Lima and channels much of its resources to regional programs. The center has established five regional potato research networks in Latin America, Africa, and Asia (PRACIPA, PRAPAC, PRECODEPA, PROCIPA, and SAPPRAD; Appendix 1).

Some international centers such as IRRI and CIP encourage publication of research results in scholarly and popular outlets. Networks coordinated by such centers also urge the dissemination of results in outside publications. Other international centers such as CIMMYT have sometimes regarded publishing in scientific journals as a drain on time and resources that could be better spent producing tangible products. Such centers see themselves as in business to produce crop protovarieties and other technologies and to train people; universities are seen as the domain of the publication scramble. Publications of networks coordinated by such centers are therefore generally confined to in-house materials.

Still other centers are particularly interested in the use of computers to assemble data bases and analyze results. Networks coordinated by the International Livestock Center for Africa, for example, emphasize assembling and sharing information.

Guiding values are a significant part of the network's culture, and they change as the network evolves. Changes in guiding values are particularly noticeable if the nature of the network shifts dramatically, as from a material exchange to a collaborative research network.

Networks should adhere faithfully to the basic values agreed to at the outset of their collaboration and should assess their guiding principles periodically to evaluate their continuing significance. If, for example, a network does not stress excellence in research among its members, the quality of research is unlikely to improve. We argue that emphasis on excellence as well as communication and participation should be explicitly fostered within the network.

Leadership

Strong and enlightened leadership is one of the most important factors influencing the performance of an organization (Kotter, 1988).

Because of the informal structure of networks, effective leadership is vital for them.

A network often emerges from the vision of a single person who eventually coordinates the network, especially during its early stages. By leadership we mean creating a vision for the network, acquiring resources to accomplish its aims, setting a realistic agenda, influencing others to rally behind the agenda, and striving energetically to realize the network's goals and vision. Leaders among network participants are usually elected as coordinators, particularly during the start-up phase.

Leadership styles span the spectrum from autocratic to participatory, but there is no evidence that one style is necessarily better than another. Enlightened leadership is the key. Many successful network coordinators have found it necessary to alter their styles as the network develops.

An effective network leader is fully aware of the capabilities and resources needed and can thus set a realistic research agenda. In an international setting, the network leader needs to know how to encourage and motivate weaker participants to complete their share of the work. Tact, diplomacy, and patience are essential ingredients for an effective network leader in international research. A network is a decentralized organizational form so the coordinator needs to operate accordingly and facilitate communication among participants, always keeping the network's goals in sight.

The coordinators of international agricultural research networks are often scientists at international centers in the Third World or universities in industrial countries with well-developed support facilities such as computers and communication links. Sometimes two international agricultural research centers team up to coordinate or plan networks. For example, IRRI and the International Fertilizer Development Center (IFDC), headquartered at Muscle Shoals, Alabama, jointly operate the twenty-one-country International Network on Soil Fertility and Sustainable Rice Farming (INSURF).

In some cases, the post of network coordinator is fully funded. In very large networks, coordination is a full-time job; international centers often coordinate the larger agricultural research networks because they have the resources to pay for a full-time coordinator. Such is the case with the IRRI-coordinated International Network on Genetic Enhancement of Rice (INGER; formerly the International Rice Testing Program), a global rice nursery in which more than fifty countries participate every year. Approximately 250,000 packets, containing information from twelve to twenty-eight nurseries, are sent out to INGER participants each year. A major undertaking of this nature, which involves numerous tasks such as collating and publishing reports, justifies a full-time coordinator.

The steady growth of another network, PRECODEPA, a regional potato research network covering Central America and the Caribbean, necessitated the establishment of a full-time coordinator in 1987. The network started in 1978 with seven countries, but by 1985 three more countries had joined. The PRECODEPA coordinator, supported by the Swiss Development Cooperation (SDC), is elected by network members for a two-year term.

In most networks, though, the coordinator typically pursues an active research program, which may be partly or wholly related to the network, and may have occasional teaching obligations outside the network. The coordinator's host institution may grant release time if the administration is convinced that the network furthers the institution's needs. In all facets of networking, then, the compliance and goodwill of directors and managers within research institutions are equally as important as the enthusiasm of coordinators and collaborating scientists.

Governance

Networks have several forms of governance. Some networks, such as the Bangkok-based International Board for Soils Research and Management (IBSRAM), emulate international agricultural research centers with a board of trustees as the ultimate authority for decision making. In other networks, steering committees set policies, and executive or program committees are responsible for seeing that policy decisions are carried out.

Four aspects of governance are important for the success of networks:

- Setting and formulating policies
- Overseeing implementation of strategies and plans
- Establishing good relationships between the governing body and the network leader
- Managing the internal affairs of the governing body

A successfully governed network does all four well. An effective network establishes and implements sound policies, nurtures working

Figure 4. Collaborators in the International Network on Genetic Enhancement of Rice planning deepwater rice improvement work. Courtesy of the International Rice Research Institute.

relationships between network leaders and other participants, and manages the joint enterprise efficiently.

Coordinators of material exchange, scientific consultation, and collaborative research networks are typically assisted by advisory, steering, or working committees. Also referred to as governing boards or boards of councillors, such advisory bodies work in conjunction with the coordinator to establish the research agenda and make midcourse corrections to the research effort. Advisory committees normally meet once a year to ascertain progress, identify bottlenecks, and chart future work (Figure 4).

A major function of advisory committees is to lighten the administrative work load of network participants and to maintain the collaborative research effort. When there are a variety of funding sources and disbursements are uneven, a typical response is to call for a meeting. But committee meetings are time-consuming and can sap productivity. By clearly outlining policy and standardizing operating procedures, advisory bodies lubricate the collective effort.

Advisory committees are usually made up of network participants, but outside experts are sometimes asked to serve. The number of people on advisory committees or boards, the way they are chosen, and their terms of service vary. The governing board of IBSRAM has seven members and soon will be expanded to eleven, elected by a committee of donors supporting the network.[1] Advisory bodies become unwieldy if the membership is too large so they rarely contain more than a dozen people. To streamline decision making, no more than seven people serve on the steering committee of the West African Farming Systems Research Network (WAFSRN), and no member country can have more than one representative on the committee. Members of WAFSRN's steering committee serve fixed terms so all seventeen nations within the network eventually have an opportunity to be represented.

Some networks are guided by more than one overseeing body. INGER, with around eight hundred collaborators, has grown so large that regional advisory committees have been set up in addition to a central committee. The INGER coordinator for Africa is posted at the International Institute of Tropical Agriculture (IITA) in Ibadan, Nigeria, and the coordinator for Latin America and the Caribbean operates out of CIAT in Cali, Colombia.

PRECODEPA has a two-tier committee structure: a regional permanent committee (COPERE) and an executive committee (COE). COPERE has ultimate authority over the network and is composed of two potato scientists from each of the ten countries participating in PRECODEPA. COPERE's tasks include nominating the executive committee, drawing up a regional program for potato improvement, establishing research priorities and appointing project leaders, seeking external funds to support the network's projects, and setting up a technical committee to oversee projects. COPERE meets once a year at the network's annual meetings to assess programs and evaluate proposals for fresh lines of inquiry. The annual budget is also approved at this time (ISNAR, 1985).

The executive committee, represented by the network coordinator, assistant coordinator, and secretary, is responsible for following up on the execution of the network's projects. COE oversees implementation

1. Donors for IBSRAM include the Australian Development Assistance Bureau (ADAB), the Australian Centre for International Agricultural Research (ACIAR), the German Agency for Technical Cooperation (GTZ), the Canadian International Development Agency (CIDA), the Overseas Development Administration (ODA) of the United Kingdom, the U.S. Agency for International Development (USAID), and the French overseas research organization (ORSTOM—Office de la Recherche Scientifique et Technique d'Outre-Mer).

of COPERE's decisions, organizes annual meetings, arranges technical reviews, and, if necessary, arranges for the hiring of consultants or other personnel for PRECODEPA. COE usually meets more than once a year.

Strategy

The conceptual framework we use to analyze networks is in many ways dependent on the network's strategy. A network strategy describes a vision of the network's future, points to the common goal, and justifies the identified path (Ozgediz, 1988). Its strategy, then, serves as a guide to the network's activities.

The three other components we have listed under guidance are extremely important in shaping the network's strategy. The strategy encompasses the guiding values of the network, incorporates broad policies adopted by its governance mechanism, and perhaps most important, reflects the vision of the network's leader.

As with any other organization, every network should have a mission statement that summarizes its strategy. The mission statement should identify the network's clients, which of the clients' needs the network intends to meet and why, what business or businesses the network should be in, the goals the network should pursue in each business, and the strategies and priorities the network should follow in meeting those goals. In addition, the strategy of the network should pay close attention to operational implications of implementing the chosen strategy. Particular attention needs to be focused on how the network should be organized, how it should acquire and manage the resources it needs for its activities, and how it should link its members.

Management of Network Resources

Networks and other organizations need inputs or resources to carry out their activities. We differentiate between four types of resources—human, financial, physical, and information—necessary for carrying out the network's activities.

Human Resources

Human resources are the most important asset of a network. Excellence in research hinges above all on the quality of scientists and technicians. Networks that attract highly productive scientists are most likely to be successful. By human resources we refer not only to the leader at the core of the network but also to the scientists who carry out research.

A network does not often have as much freedom to select its members as does a freestanding organization. Strong political or economic pressure sometimes dictates the choice of network participants. In view of such exigencies, network leaders and other members may need to coax the less competent participants along.

Acquiring human resources is only one facet of managing a network's human resources. Once people have joined a network, productive individuals need to be retained and further developed. It is easier to retain high-quality participants when they are motivated and excited by their involvement in the network. Nurturing the scientific and technical capabilities of members by the coordinator and others is extremely important. Many networks have found that training programs, monitoring tours, and workshops are essential to foster the members' commitments to the network's guiding values and mission.

Training courses upgrade the skills of network participants and help create collegial links among members. Training enhances the scientific capacity of present and future participants, thereby helping make the joint effort more effective. Also, while taking courses trainees make friends and valuable contacts to whom they often turn later for assistance (Figure 5). Such amicable relationships greatly facilitate the exchange of information and technology between national programs; training courses thus reinforce the rim concept (see Chart 2) in networking.

Many material exchange and research networks offer training courses specific to the needs of their programs. Such courses, which operate from a few weeks to several months, provide a mix of classroom, laboratory, and field experience. During the first six years of PRECODEPA, 110 people attended the network's courses and workshops (ISNAR, 1985:9). The Southeast Asian Program for Potato Research and Development, a sister network of PRECODEPA linking potato researchers in the Philippines, Indonesia, Thailand, Papua New Guinea, and Sri Lanka, sponsored three courses in 1986. Two courses conducted in Jambegede and Lembang on Java focused on technology transfer by extensionists. Instruction was in English and Bahasa Indonesian (CIP, 1985b). The third course, on regional seed production, ran for six days at Baguio in the Philippines.

Another jointly planned research effort that strongly emphasizes

Figure 5. Participants in PRAPAC, a collaborative research network in the Great Lakes Region of Africa, attending a network course on germplasm improvement in Rwanda. Courtesy of the International Potato Center.

training is the Trypanotolerant Livestock Network, which is coordinated by ILCA. This network investigates the epidemiology and control of a major livestock disease and operates an annual seven-week course in Nairobi with the help of the International Laboratory for Research on Animal Diseases and the International Center of Insect

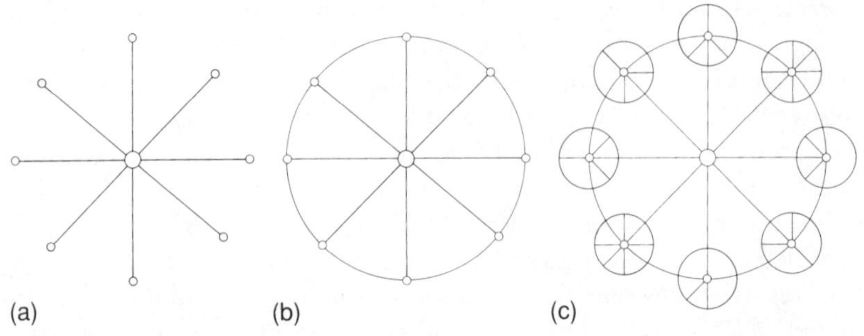

Chart 2. Schematic view of network linkages: (a) hub-and-spoke networks in which information radiates largely from a hub to participants, (b) the rim effect in which participants collaborate with each other as well as with the hub, and (c) subnetworks at the national level generated by an international network

Physiology and Ecology (ICIPE). The Trypanotolerant Livestock Network course, which began in 1982, is offered in English and French (ILRAD, 1986a). By the end of 1985, forty-two network participants had taken the course.

The Regional Network for Agricultural Machinery (RNAM; Appendix 1) devotes 37 percent of its budget to training; from 1978 to 1986 it provided fellowships to ninety-five individuals from eight countries (RNAM, 1986). RNAM fellows have traveled to various institutions to receive further training in machinery design, testing, and evaluation, as well as manufacturing technology and industrial extension. These numbers may not be massive, but properly trained people at key nodes in a network are vital to the ultimate success of the collaborative effort. Industrial extension is one of the weakest links in the multistep process of developing and delivering machinery for Third World farmers, and RNAM has wisely supported training in this area.

Material exchange, scientific consultation, and collaborative research networks generally organize regular monitoring tours to assess progress and problems at research sites. Sometimes referred to as study tours to avoid any connotation of checking up on participants, such tours are composed mostly of collaborators from national programs, the main actors in networks. Monitoring tours are an integral part of collaborative research ventures but are somewhat less frequent in other types of networks. Monitoring tours are important mechanisms for building expertise among network members.

Monitoring tours provide opportunities for participants to discuss ideas and obstacles at length with fellow participants (Figure 6), and they place participants in the field, where they can see difficulties and successes firsthand. Insights and inspiration ensue from well-conducted monitoring tours.

Germplasm exchange and testing networks employ monitoring tours to demonstrate the value of collaboration to potential participants and to explore problems and prospects for existing collaborators. INGER, a material exchange network, organizes two to three monitoring tours to several countries each year. Each tour contains twelve to eighteen participants and focuses on a particular rice type or problem such as deepwater rice or rice germplasm for acid soils.

Workshops are an integral part of research networks and an important forum for developing human resources. Workshops, or "networkshops" as they are sometimes called (FSSP, 1986), can serve as annual meetings or may be held in addition to them. Workshops usually take place annually but may be organized at any time to discuss

Figure 6. Collaborators in the Asian Rice Farming Systems Network on a monitoring tour in Bangladesh. Courtesy of the International Rice Research Institute.

an emerging problem. As do monitoring tours, workshops help cement the professional bond between network collaborators, thereby promoting the rim effect. Workshops provide an arena in which to assess difficulties, analyze results, and make appropriate adjustments to the research effort.

Financial Resources

Networks are sometimes not recognized as full-fledged organizations so they may encounter difficulties in acquiring and maintaining financial support. One of the main advantages of networks, the relative ease with which they may be established or ended, at times can work to their disadvantage. A government or group of donors might think twice about closing down an institution, but dissolving a network is less traumatic. Prudent financial management is thus especially important for the survival of a network.

Five aspects of financial management impinge heavily on the success of a network: acquiring funds, allocation of resources, financial management, financial control, and financial systems and procedures.

Networks need funds for the coordinating unit and the nodes. The coordinating unit needs an adequate budget to begin the network and carry out its activities. Start-up and maintenance funds usually come from a committed donor or from the institution housing the coordinating unit. It is extremely important for the network's core or hub to have stable funding. The coordinator thus needs to hone grant-raising skills so that the network is continually infused with funds. Acquiring funds is much easier when a network can communicate results or successes clearly and expeditiously.

Although primary responsibility for fund-raising usually rests with the coordinator, other network members should be involved in approaching donors. Fund-raising should not be envisaged as the sole responsibility of the coordinator. Furthermore, the institution housing the coordinator needs to be informed periodically of the network's activities and successes to forestall withdrawal of its support for core activities.

Financing the activities of the members of the network often presents a different story. Several models or financing modes are in operation. The host institution of a network node usually provides the funding necessary for that member's participation. Most network participants contribute in kind by devoting a certain percentage of their time to network tasks. Resource-poor national agricultural research systems, however, often need some bilateral support to encourage network activities. In the case of the Programme Regional d'Amelioration de la Culture de Pomme de Terre en Afrique Centrale (PRAPAC), a potato research network in the Great Lakes region of Africa, local missions of the USAID in Burundi, Rwanda, and Uganda support the network's activities in their respective countries. Alternatively, an outside donor may underwrite the entire cost of the network, but this rarely happens because participants are likely to be "bought," rather than join because of innate interest in the scientific merits of the joint enterprise, a point we will amplify in our discussion of network principles.

To the extent possible, networks should attempt to avoid dependence on external funding but strive to finance their activities through their own channels. External financing of networks should be viewed as a short-term measure to get the network off the ground. Chronic dependence on external funding may render the network unsustainable because donors generally prefer to play catalytic roles and are wary of continually pouring funds into loose-knit organizations such as networks.

A network would be in a privileged position if it could finance all its activities through unrestricted sources because it could then use its financial resources more efficiently. When network activities are funded by restricted contributions, such as for special projects, deviations from the network's strategic direction may occur. When unrestricted funding is not available, network coordinators and participants should make an effort to ensure that the activities supported by restricted funds are fully complementary with the network's strategy.

Sound financial planning is essential. Financial planning cannot be undertaken in isolation from program planning; therefore, financial managers of a network must see their primary role as transforming program plans into financial terms. Each network member needs to plan his or her activities within the prevailing financial constraints and budgetary procedures. Financial and program planning are often carried out by the coordinator to provide an integrated financial plan for the entire network so that a coherent, overall budget for the network's activities can be submitted to funding agencies.

Financial control of material exchange, scientific consultation, and collaborative research networks is normally exercised through internal and external auditing procedures. Networks organized along the lines of an international center are subject to external and internal audits. External audits help ensure that the network's financial statements accurately portray its financial standing. In addition, external auditors often make suggestions for improving financial controls or policies. Sound external audits confer greater financial credibility to a network, an important consideration for donors.

The final aspect of financial resources we wish to touch on is financial systems and procedures. Networks that are not organized as independent entities normally depend on the existing financial systems and procedures of the host institution. In such cases, financial systems for accounting, budgeting, reporting, financial analysis, and cash and currency management are performed by the finance department of the host institution. If the host institution's financial systems and procedures are sound, the network will benefit. But if each of the network members follows different financial practices, such as in Anglophone and Francophone countries, reconciling these accounting approaches is an added burden for the network coordinator.

Problems associated with financial systems can be avoided if the basic spirit of a decentralized operation is followed. Participants should be encouraged, rather than instructed, to improve their accounting practices. No attempt should be made to control the activities of individual participants. Imposing the system employed by one participant on the entire network can be counterproductive and is often not feasible.

Network coordinators should exercise extreme caution in devising financial systems. The aim should be to acquire the most essential financial information for the purposes of decision making. Insisting on too much data can lead to unnecessary bureaucratization and red tape.

Apart from adequate amounts of funding, flexibility in disbursement is essential. Too many special project funds, each with its own disbursement schedule, can interrupt the smooth functioning of a network. Special provision should be made for bridging funds in the event that disbursements from donors do not synchronize with the network's activities and funding shortfalls result. A sympathetic attitude on the part of host institutions can help fill temporary funding gaps. The finance department of ILCA, for example, occasionally steps in to meet temporary funding lapses of the Trypanotolerant Livestock Network. Host institutions are willing to extend credit only to effective networks that enjoy the confidence of donors.

Physical Resources

By their nature, the physical resources of a network are spatially separated. In some cases, a network's physical resources are located on different continents and on widely scattered islands. Particular attention needs to be paid to managing those resources most effectively for the common good. The same caveat applies to other organizations in which research, development, or manufacturing are carried out at a variety of sites.

Under physical resources we include research stations, specialized equipment, laboratories, buildings, procurement services, and general services such as housing, food services, transport, and security. To be effective, network participants need to have access at the appropriate times to all the physical resources necessary for their work.

The quality and availability of physical resources vary among network participants. Improvement of the physical resource base is usually outside the scope of a network. Government procurement systems, for example, dictate how buildings or other facilities for a research station are handled. In some cases, new buildings such as a glasshouse or training facility may be constructed for a network, but network members usually must operate with existing facilities. Under these circumstances, they need to be aware of each others' physical resources and plan their activities accordingly. Again, ingenious orchestration by the network's coordinator is called for.

Information Resources

Information, like human, physical, and financial assets, should be treated as a resource. For information exchange networks, information is one of the most important resources and represents the very lifeblood of the network. The generation and exchange of information is the main business of such networks. Information in networks is generated, collated, and then disseminated in usable form to participants and outsiders. The process of transforming information requires several steps, and each stage should be carried out as effectively and efficiently as possible.

Each node of the network carries equal responsibility for acquiring, processing, managing, and disseminating information. The coordinator has the major responsibility for circulating necessary data for other members.

A network's capacity to manage information can be assessed by examining its effectiveness in computing, office automation, record keeping, and library and documentation services. External donors could provide bilateral funding to individual members to help them upgrade their information management capacities. With enhanced capabilities to receive and analyze information, each member is able to contribute effectively to the network's activities.

Successful networks incorporate information services and technological aspects of information processing in their strategic plans. A network should first develop a strategy and examine its implications for information management at each site. If necessary, measures should be taken to raise substandard sites to a minimum desired level of quality.

All networks produce publications to inform participants about research results and new developments. Some networks operate essentially as information conduits with the hub periodically sending newsletters or abstracts to individuals on a mailing list. Material exchange, scientific consultation, and collaborative research networks typically provide a range of publications to inform participants about the outcome of trials and the emergence of new designs, procedures, and technologies.

The newsletter is the most common form of publication employed by networks. Newsletters contain such items as announcements of upcoming meetings, training courses, and recently published articles and books relevant to the network's focus. International nursery networks generally publish annual reports containing the results of various trials. To help participants obtain information useful for planning nurseries, INGER also distributes interim reports containing an analysis of preliminary returns from nurseries.

Some networks also publish the results of workshops and annual conferences. To reduce costs and speed production, most network publications are either mimeographed or computer-generated. Networks in developed countries are able to access electronic bulletin boards from workstations and computer terminals, thereby accelerating the transfer of information among participants. In the final chapter, we will have more to say about the role of information technology in networking.

Information management is also important in generating data for decision makers, particularly the coordinator, policy and executive committees, and other managers. The network coordinator can play a useful role in encouraging the development of appropriate management information systems for the network.

Networks with their hub located in institutions with strengths in information management are likely to be most successful in managing their information resources. It is no accident that several successful networks have originated at ILCA, which has a deserved reputation for effective information management (Appendix 1).

Information technology and management are increasingly recognized as devices for integrating organizations. This cohesive function of information management is particularly important for networks that have greater needs for integration than freestanding organizations located at a single site.

Management of Network Activities

The work side of networking includes a cluster of four factors: operational plans, control systems, coordination and communication structure, and work processes (Chart 1).

Operational Plans

Networks, like other organizations, typically undertake an integrated planning process (Chart 3). Integrated planning normally has

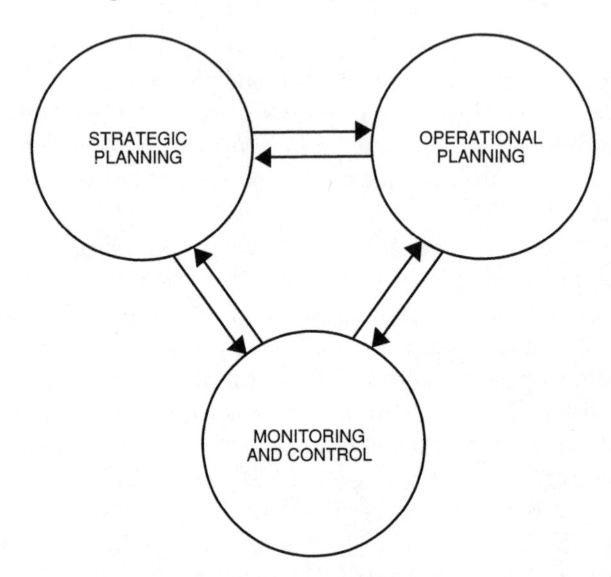

Chart 3. Integrated planning process for networks and other organizations

three stages: strategic planning, operational planning, and control and review mechanisms. Strategy formulation was explored in our earlier discussion of network guidance. We now address the other two components of the integrated planning process.

Operational planning involves making the network's strategy operational. This requires translation of the main direction of the network into specific activities and projects that are needed to accomplish the network's mission. Successful networks take planning seriously and prepare medium-term—typically for three to five years—as well as short-term plans, usually for one year. The plans reflect the priorities articulated in the network's strategy and cover activities to be carried out by all members of the network.

Budgeting is an important component of operational planning. The costs of the various network activities are estimated and put in a program and budget framework for a specific period. Planning and budgeting exercises require participation by key network members, usually a policy or steering committee.

Successful networks adopt realistic plans. It is far better to work toward a few achievable goals than to stretch too far. A network may evolve eventually into a larger, more complex enterprise, but it is better to start from a modest but realizable base. When plans are implemented because they are realistic, participants are encouraged. When attempts to implement overambitious plans fail, network members rapidly become disillusioned and the network may stall.

Each network node should ideally have an operational plan of its own that links with the network's overall operational plan. This individual or nodal plan could be in the form of project plans or simply cover all activities to be carried out by that particular node. A broad network plan that incorporates all nodal plans could increase the network members' sense of ownership.

Control Systems

The third phase of the integrated planning process involves reviewing and monitoring performance that we refer to as control systems. Each network should have a mechanism for assessing the outcomes of implementing its strategy and plans. Good strategic planning requires sensitivity to the shifting contours of the network's environment. A network should be able to revise its strategies periodically in response to emerging issues.

Review activities of networks take several forms. Some networks formed as freestanding institutions often go through external reviews organized by their donors. Regardless of whether there is an external review, each network should have a sound internal review process. In addition, the governance mechanism of the network plays an important oversight function. Monitoring tours, workshops, and annual meetings also serve as valuable review devices.

Each participating network member should have a local review procedure to assess how well it is performing. The results of such in-house reviews should be shared with other network members. Sharing this knowledge may lead to an exchange of views on how a particular member could further improve his or her operations.

Coordination and Communication Structure

All networks are anchored to a hub where work is coordinated and research results are collated. The location of a network hub may change from time to time, but there is always a need for a nerve center to collate and process information and technology. The coordinator is usually elected by members of the network and may retain his or her post for the life of the network. Sometimes this position is rotated among network members to provide leadership training and to share the task of administration.

Several mechanisms are used for coordinating the network's activities. Coordinators, policy or steering committees, executive committees, workshops, publications, and monitoring tours all help give the network direction and cohesion (Table 2). Simple networks, such as information exchange networks, rarely have all these mechanisms for furthering their goals, but collaborative research networks usually need all these devices for planning and fostering close communication among participants.

Perhaps the feature that most distinguishes a network from other organizational forms is its decentralized structure. Typically, the structure of an organization depicts formal reporting relationships among units, illustrates how the organization differentiates among the tasks and activities carried out by its members, and shows how the activities of different units, such as research, manufacturing, and marketing, are to be integrated and coordinated.

Differentiation of tasks, projects, and activities is usually done by the network members by taking into account the network's mission and the comparative advantage and capabilities of each of its participants. Tasks should be assigned so as to minimize duplication, maximize communication among units, and achieve a high degree of efficiency. But the voluntary nature of most networks leads to some overlap, duplication, and inefficiency.

Each organization has a communication structure that shows how different individuals and units interact with each other. Although networks may seem to be amorphous, they exhibit structures and features that can hinder or enhance internal communication. Here we briefly explore a hub-and-spoke model, consisting of a core or hub, lines of communication radiating from the coordinator, and interactions among participants (Chart 2).

Table 2. Mechanisms for improving planning, management, and communication within networks

	Network types			
Mechanism	Information exchange	Material exchange	Scientific consultation	Collaborative research
Coordinator	X	X	X	X
Policy committee			X	X
Executive committee			X	X
Workshops			X	X
Training		X	X	X
Monitoring tours		X	X	X
Publishing program	X	X	X	X

The simplest networks operate in a hub-and-spoke arrangement, in which most of the interchange takes place between the coordinator and the widely scattered participants (Chart 2a). Information exchange networks operate in this manner. Interchange is limited to ideas and data and is essentially unidirectional: from the hub to subscribers. A limited amount of information is transmitted to the coordinator from participants.

Networks in which participants interact with each other, as well as the coordinator, generally display the rim effect (Chart 2b). Peripheral linkage, pulling together widely scattered researchers so that they learn from each other, is one of the most valuable benefits of effective networking. Peripheral linkage is particularly evident in material exchange, scientific consultation, and collaborative research networks.

The benefits of the rim effect will be explored in more detail in later chapters, but an example here will help clarify the concept. With support from the United Nations Development Program, the governments of Australia and Japan, and ten collaborating countries in Asia, the Regional Network for Agricultural Machinery links more than 230 collaborators in the design and testing of agricultural machinery. Since RNAM's inception in 1977, more than fifty different machine prototypes have been exchanged among collaborators in Bangladesh, India, Indonesia, Iran, Nepal, Pakistan, Philippines, South Korea, Sri Lanka, and Thailand. RNAM pays for international shipping of machinery, and UNDP arranges for its duty-free passage through customs. Recipients must pay for the cost of making the prototype. Only a few of the prototypes are suitable without modification in the recipient country; most of the arrivals must be adapted for local use, and some are found to be unsuitable. Even discards save the recipient time and money because the ineffective design was developed elsewhere. Furthermore, a failure in one country might prove to be a success in another.

The rim effect can also be envisaged as horizontal technical transfer, or lateral exchange, between participants (CIP, 1985a). Whichever visual image is most appropriate, the underlying idea is that all partners should have equal opportunity to participate in decision making and planning. The coordinating institution should not be envisaged as being on a higher plane, dictating to a lower echelon of network members. Networks are coordinated, not controlled (Lipnack and Stamps, 1987). Effective networks are not set up like tripods to raise one part of the structure so that it dominates the other components.

In addition to the rim effect, some radial nodes spawn subnetworks, often at the national or regional level, to tackle specific problems (Chart 2c). This is often the case with large, organizationally complex networks, such as international nurseries that screen crop germplasm for various attributes, or multipurpose commodity networks. The subnetworks are still tied into the larger network for fresh germplasm or other technologies. The transfer of research methodologies and testing procedures is facilitated if a subnetwork remains linked with its larger, parent network.

Although network members are linked through various communication channels, the decentralized, nonauthoritarian nature of networks means that members do not have to abide by any formal reporting regulations to the network hub. The network leader must rely on persuasion rather than coercion to extract results from participants. Highly motivated participants in successful networks usually provide results and other information on a regular basis.

Ideally, research networks are composed of autonomous agencies or organizations working under self-management rather than under a hierarchical authority (Litwak and Hylton, 1962). Decentralization, exemplified by the rim effect, is a key ingredient in successful networking (Van de Ven and Ferry, 1980). In a totally decentralized network all agencies participate equally and symmetry or reciprocity characterizes flow of information and resources among participants. Although few networks are perfectly symmetrical in this regard, decentralization is a key ingredient in successful networking.

High-profile international networks sometimes spawn subnetworks of research sites at the national level (Chart 2c). Enthusiasm generated by the sixteen-nation Asian Rice Farming Systems Network, for example, has spurred the Philippines national program to establish 115 cropping systems research sites across the archipelago country (V. Carangal, pers. comm.)

Work Processes

Work processes include data collection, analysis, testing, fieldwork, and similar activities carried out during research and experimentation. From an organizational standpoint it is important that the network and its members identify the right processes for carrying out the work and implement them effectively. Work processes are described in the chapters dealing with information exchange, material exchange, scientific consultation, and collaborative research networks.

Management Skills and Teamwork

Management skills and teamwork refer to the management practices used by managers and supervisors in the network and the presence or absence of a team spirit among the participants. Both factors are difficult to quantify, yet they are important to the success of a network. Each network has a number of individuals who play managerial or supervisory roles. The skill of these people is reflected in the network's efficiency and effectiveness.

By management skills we mean the skills of the individual manager in goal setting and work planning, organizing and coordinating, directing and delegating, supporting the work of subordinates and problem solving, reviewing and providing feedback, and motivating and communication.

The teamwork concept is more difficult to define. Because of its decentralized and voluntary nature, a network needs a widely shared spirit to promote productive intragroup relations. The structure and operating procedures of a network and its components can facilitate, or hinder, cooperation and teamwork. The role of the network leader is extremely important in this regard. A top-down, control-oriented leader may run counter to the participatory team spirit that most networks need. Successful network leaders are persuasive and care about the development of the team. Cooperative activities carried out by the network should be examined carefully to assess their potential to increase the team spirit.

Network Environment

Four main groups of stakeholders in the network's environment greatly influence its conduct and culture (Chart 1). The four main outside stakeholders are clients, donors, host institutions, and other institutions and networks. To be effective, a network needs to manage properly its relationships with each of these key stakeholders.

Clients

As with any organization, clients are a network's most significant stakeholders. Clients include individuals or institutions that benefit directly from the products or services provided by the network. Although the members of the network are in a sense also its clients, here we refer to outside individuals and institutions who receive services from the network.

The clients of networks organized as international centers are typically defined as the national agricultural research systems in a specific geographic region. For other networks, national research institutions in the participating countries are the key clients. For effective relationships with clients, network members need to be atuned to their present and future needs. Appreciation for clients' needs is gained through frequent contact with them so that products and services can be tailored to their needs. Each member of the network should serve as a vehicle for transferring information about clients to the leadership of the network. Workshops that include clients provide an opportunity to assess their needs.

A network should strive to establish a strong orientation to clients from the outset. The only way to decipher clients' needs is to maintain constant contact with them. Network members should be especially careful to represent a broader perspective than their narrow interests or personal biases.

Donors

Networks need to court and actively manage their relationships with donors. Although these relations are primarily the responsibility of the coordinator and the policy committee, each network member should initiate and maintain positive interactions with donors.

Some international agricultural centers have established a small donor secretariat as part of the office of the center's director general. Even though it may not be possible for a network to establish such a secretariat, a similar albeit more modest arrangement to manage donor relations is desirable. Donors themselves may establish coordinating groups to encourage networking. For example, SPAAR has a special Networking Working Group to deal with networking in African agricultural research.

Donors are typically interested in the overall impact of a network so it is important to keep them informed at all times of developments in a network and especially of its major achievements. A newsletter summarizing activities is often a useful communication device. Virtually all networks publish a newsletter, even if it is a simple mimeographed release. Frequent personal visits to key donor agencies may be necessary to keep the interest of the network's financial supporters and to make them aware of the network's progress. Such visits also help network members gauge future funding prospects.

Interactions with donors also help shape the scope and direction of a network. Because of the crucial importance of donors, particularly during the early stage of a network, they can greatly influence the culture or characteristics of a network. Former colonial powers, for example, tend to provide support for networking in their former colonies, often in some competition with, or in isolation from, each other. Australia is most interested in fostering networks in its geopolitical theater in the Pacific and Southeast Asia. Major donors may use leverage in the selection of a network's coordinator or headquarters.

Successful organizations, including networks, find ways to manipulate their environment to minimize threats and constraints and to take advantage of opportunities. Identifying priorities of major donors and tailoring proposals to funding agencies is one way networks can further their aims. Testing political winds and identifying key "buzzwords" that ignite the interest of donors are important skills for network participants to ensure the continued support of their collaborative efforts.

Networks may even change their names to reflect shifts in direction and to be more in tune with donors' concerns. The International Network on Soil Fertility and Sustainable Rice Farming (INSURF; Appendix 1) was formerly known as the International Network on Soil Fertility and Fertilizer Evaluation for Rice (INSFFER). IRRI-coordinated INSFFER was always concerned with sustainable soil management practices, but the latest change in name made this underlying concern more explicit and is in line with donors' concerns about sustainable agriculture. Maintaining and enhancing strategic alliances with donors and other players in the international agricultural research scene are crucial to the survival of a network.

Host Institutions

Host institutions are the organizations within which the core and participating members of a network are situated. Although one can argue that host institutions should be considered a part of the network, we prefer to regard them as separate parts of the network's environment. The leadership of the host institution should be constantly kept abreast of the network's activities and future plans. Network members should continually seek support for the network within their own institutions. Often this involves sharing information, but natural rivalries in any institution are likely to involve network members in internal problem solving, firefighting, negotiation, and conflict resolution. This again is an area in which good management skills among network members are helpful.

Internal policies and procedures of host institutions often pose problems for networks. The host institution's system for recruiting staff, purchasing materials, financial management, or use of computers may not fit the needs of the network. In such cases, network participants need to adjust their styles accordingly, rather than try to change the policies and procedures of the host institution.

Other Institutions and Networks

To carry out its mission, the network may need to collaborate with other institutions or networks. Outside institutions are mostly organizations in the host countries of network members (other than the host institution), advanced research institutes in developed countries, organizations providing information and other services to the network, and international organizations.

Managing relationships with these other institutions is no different than catering to donors and clients. The network should actively scan its environment to detect changes in these institutions and adjust its program accordingly. The network's strategic plan should explicitly recognize the institutions with which it expects to collaborate and share responsibilities.

Interinstitutional committees, placing liaison persons in other organizations, and workshops and symposia are common devices for fostering relationships between networks and outside agencies. The network's communication and publication arm plays a crucial role in keeping all external stakeholders informed of the network's activities.

Tenets for Effective Management of Networks

Some eighteen major tenets or principles can be distilled from our review of network planning, management, and communication (Table 3). These tenets reflect facets of our conceptual model (Chart 1). A network does not have to abide by all of these tenets to function smoothly, but the more items on the checklist that apply, the greater the chances for operational efficiency and network success.

Table 3. Tenets for effective planning and management of networks

1. Widely shared values
2. Effective governance and policy-making structure
3. Effective coordinator/leader
4. Clear, well-focused strategy
5. Decentralized coordination and communication
6. Sound, pragmatic work plans
7. Effective review mechanisms
8. Appropriate work processes
9. Good system for recruiting and keeping network staff
10. Stable and adequate funding
11. Adequate physical resources
12. Effective information management and sharing
13. Good linkages with clients
14. Productive dialogue with donors
15. Healthy relationships with host institutions
16. Effective ties with other networks and institutions
17. Skilled research managers at the network nodes
18. Effective teamwork among participants

Organizational Advantages of Networks

A standout feature of the organization of most networks is their relative simplicity and flexibility. This characteristic often results in more effective use of existing but underused research capacity. For the most part, networks take advantage of existing facilities and staff rather than erecting new buildings and adding personnel. Because no major investments are generally made in setting up networks, they can easily and painlessly be dismantled when their job is done, funding is not renewed, or they have failed. Unlike large bureaucracies, which tend to feed themselves and remain in existence because of inertia long after they have outlived their usefulness, many networks are, and should be, ephemeral.

Another way that networks maximize results from research expenditures is by making better use of existing information than can scientists working in isolation. Scientists often argue that information is insufficient and that more research is required, particularly when they apply for research grants. Conversely, administrators and donors sometimes complain that researchers produce stacks of data and reports, much of it inaccessible and in need of more analysis.

Our knowledge base on many issues in agricultural research is deficient, and more research is certainly needed. Most networks are rightly set up to expand the frontiers of knowledge, but one of their benefits is that existing information is more widely shared and thus has a better chance of being used. All four types of networks facilitate information dissemination through publications and other media.

Networks not only capitalize on existing knowledge, but they help prevent redundancy in research. Reinventing the wheel is one of the dangers of individuals or groups working in isolation. All four network types help reduce unnecessary overlaps in research, but collaborative research networks are especially useful in this regard because the research agenda is jointly planned from the start.

4

Development Stages

Each network has its own development path, which can be evolutionary, arising from spontaneous needs and shifting priorities, or which can be planned and set in motion from the outset. We do not suggest that there is any formula for starting or running a successful network, but some concepts and experiences gained from ongoing networks can help in initiating and running new ventures. Ideas and generalizations about network development can apply to networks that focus on problems requiring applied research as well as sophisticated collaborative research efforts at the leading edges of science.

Networks may start with one purpose and end with another. Some may be circumscribed in scope with a small membership, while others may have multiple research goals and involve hundreds of participants. Regardless of the research focus and number of collaborators, to remain vital networks must respond to changing needs of participants as well as keep abreast of new research methodologies and technologies.

Here we characterize network factors that change over time such as articulation, differentiation, evolution in type, degree of formality, and degree of centralization. We then examine four stages of a network from initiation through early growth and maturation to dissolution.

The Concept of a Development Pathway

As we emphasized in Chapter 3, each network should have a strategy and a plan to guide its work. The plan should spell out the network's goal, identify clients, and explain how plans are to be implemented and managed and how the network will be financed. All of these factors must be considered in charting the development pathway.

Material exchange, scientific consultation, and collaborative research networks should have strategic plans that lay out their anticipated trajectories. Such plans should include an analysis of the problem to be addressed, the rationale for a network approach, a vision for the network and its partners, ways to achieve desired goals, and a strategy to move from the present situation to some desired future improvement.

An authoritative founding document that analyzes the problem to be addressed is an important ingredient in the strategic planning process. A founding document is essential for collaborative research networks and highly desirable for research consultation networks. Information or material exchange networks may not need founding documents, except when they depend on common methodologies or procedures.

A network's strategic plan should be dynamic and provide flexible guidelines for potential development paths. As the network develops, it will modify early plans to suit current needs and accommodate unanticipated situations.

Factors Affecting Network Development

Five main features characterize network development. Factors influencing the maturation of a network include articulation, differentiation, evolution in type, degree of formalization, and degree of centralization.

Articulation

Articulation can be defined as the type, methods, and scope of relationships that develop among partners in a network. Articulation necessarily becomes more complex as research collaboration and interactions among participants increase and therefore should be planned as systematically as possible.

In the beginning of a network, articulation may consist mostly of relations between the hub and the nodes; in practice this usually consists of contacts between the network coordinator and individual scientists. As the network develops and planning becomes more par-

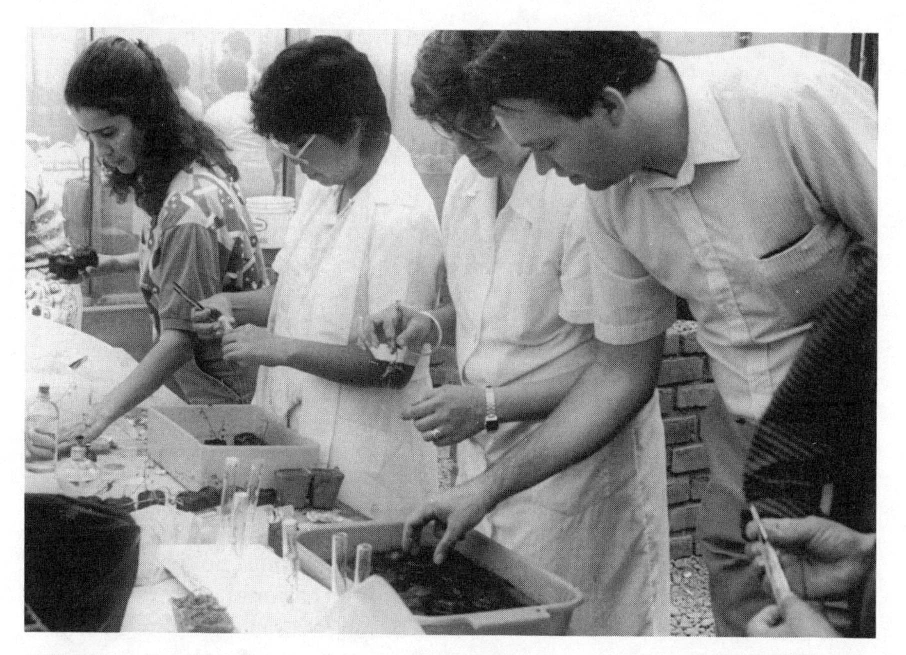

Figure 7. A tissue culture training course in Cuba organized by PRECODEPA. Training courses provide important opportunities for network participants to communicate with each other and upgrade their skills. Courtesy of the International Potato Center.

ticipatory, scientists begin to communicate directly with one another as well as with the coordinator. At that point, nodes may begin to organize specialized workshops or training courses in which scientists, both inside and outside the network, are asked to participate (Figure 7).

In collaborative research networks, articulation among scientists is especially well developed, far surpassing correspondence and the exchange of ideas at professional meetings. Articulation in collaborative research networks includes short visits to other scientists' research plots or laboratories, extended stays at each others' research institutions, and individual consultancies by scientists with special skills to other research centers in the network.

Differentiation

Differentiation can be defined as the distinctions in responsibilities between participants in a network. Differentiation results in a division of tasks and, like articulation, becomes more pronounced and complex as the network matures.

Examples of differentiation are the individual responsibilities assigned to or assumed by countries participating in the regional potato networks assisted by CIP. In regional potato research networks, such as PRECODEPA, PRAPAC, and PRACIPA (Appendix 1), network participants identify major constraints to potato production in a particular region. Each country then assumes specific responsibilities for a problem or set of constraints in which it is particularly interested and in which it has the necessary expertise.

Evolution in Type

As we pointed out in Chapter 2, networks may evolve in complexity or purpose. By evolution in type we mean the transition from one network type to another, such as from a material exchange to a scientific consultation network. The International Soybean Program (INTSOY), for example, started as an international nursery (material exchange network) in 1973 but dropped that function in 1985. INTSOY has now become a multipurpose, scientific consultation network focusing on soybean processing and utilization. The Trypanotolerant Livestock Network started informally as a scientific consultation network in 1981 but by 1984 had evolved into a collaborative research network.

Such evolutionary changes generally involve more frequent and intense interactions between participants and necessitate more management inputs. Each step along the spectrum from information exchange to collaborative research networks implies a greater commitment of national resources, a greater reliance on the collective efforts of network participants, more skilled coordination, and greater costs (Winkelmann, 1987). If such evolutionary changes are anticipated, the need for greater administrative and management inputs can be accommodated in the long-range plans of the network.

Degree of Formalization

Networks often become more formal as work proceeds. Some networks remain informal and require no signed agreements to further their purposes. As networks become more complex, requiring more articulation and differentiation, devices such as memorandums of understanding may be required. In the case of collaborative research networks, which may result in patentable or marketable products, and in which commitment of resources by participants is greater, protocols are often needed to spell out the rights and obligations of each participant.

Degree of Centralization

As with the other factors, degree of centralization is influenced by the age and complexity of the network. Networks that are new or relatively small with modest programs usually require centralized planning and control. As participants take on more responsibilities, however, and as articulation and differentiation increase, greater decentralization helps ensure the smooth functioning of the network. As participants take on more responsibility in planning and managing the network, a sense of ownership emerges. Greater decentralization of decision making also provides more opportunities to start subnetworks at the radial nodes (Chart 2c).

Stages of Development

In addition to factors that bring about change in networks, one can discern various stages through which a network may pass. Here we focus on four major stages: initiation, early growth, maturity, and dissolution. Sometimes passage through these stages is rapid, whereas in other cases networks show no sign of slowdown or dissolution after more than two decades. By their very nature and purpose, though, most networks are not permanent institutions. Dissolution does not necessarily mean failure but can signify that the mission has been accomplished.

Initiation

Almost all successful agricultural research networks have, or have had, an instigator or promoter. This catalytic role has been played by individuals working mostly alone or by international agricultural research centers, universities, foreign aid organizations, and private foundations. Whether the initiative to start a network comes from an individual operating alone or from an institution, someone must invest time, thought, and financial resources to stimulate its development. One incentive for an early commitment of effort and resources by a network instigator is that the individual or organization will have a major influence on the values and culture of the network.

A network instigator usually helps prepare a founding document and organizes the initial workshop or conference that brings potential network participants together. Such start-up procedures involve exploring dimensions of the problem to be addressed, agreeing on a venue and agenda for the first meeting, obtaining funds for the meeting and preparation of the founding document, and identifying a coordinator and a policy or steering committee.

The task of founding a network is difficult if the instigating individual lacks institutional support. An individual acting alone can promote the idea of a network, but a network may never develop if participants cannot secure support from their host institutions or donors. Although networks usually involve linkages among individuals, most scientists operate within bureaucracies and depend on them for salaries, supplies, travel funds, and support staff. For a network to have a chance for success, then, participants must obtain the scientific and financial backing of institutions, particularly their own. An individual can, of course, start an information exchange network with very modest resources since such networks do not require a founding document or initial workshop. But for material exchange, scientific consultation, and especially collaborative research networks, instigators must address the needs and concerns of research institutions and potential funding sources, as well as those of scientists.

In many cases, the instigating individual or institution may become the network coordinator. That instigators often serve as coordinators reflects their particular interest in the problem to be addressed, their knowledge and influence with respect to funding sources, and their ability to locate key individuals who might participate effectively in the network. How long the instigator remains as coordinator depends largely on how long it takes for the network to diversify its goals and become more decentralized. If the initial coordinator displays strong yet sensitive leadership, the network will likely benefit greatly and participants will be satisfied. As the network matures and members gain more capabilities, the network initiator will find it necessary to loosen the reins and create more opportunities for participatory management of the network.

A number of factors are involved in the establishment of a network. Networks are started for many reasons, but most primary purposes of agricultural research networks fall into three main categories:

- Solving important problems that are international, regional, or, for very large countries, national in scope

- Building individual scientific or technological capabilities in a given field
- Building institutional capacities in research or technology

In carrying out these three major purposes, networks bring together partners with similar or sometimes uneven capabilities. Networks linking participants with similar capabilities are peer-based associations, whereas networks encompassing individuals of unequal capabilities involve nonpeer relationships in which some members are clearly stronger and carry more weight than others. The development of a network is thus heavily influenced by the degree to which participants can be considered peers. For example, to solve a pressing international problem, a group of well-trained scientists from several countries may band together to work on that problem through jointly planned research. In such a network, it is the desire to work with peers that brings the group together, and the peer relationship is the driving force behind the network's strategy and operation. By contrast, a major purpose of networks with partners of unequal capabilities may be to strengthen them.

All networks usually require a strong central core group, which is often based on a strong research institution that provides a critical mass and scientific integrity, ensures strong financial management, and is willing to assist in coordination. This approach has been used successfully for networks that lack true peer relationships. Other core groups can be made up of senior individuals from participating countries or institutions. Scientific core groups also often supply an individual or individuals to help coordinate the early activities of a network.

The decision to assemble a network is often an outgrowth of contacts made during a professional meeting. A conference or symposium provides an opportunity for sketching the basic outlines of a proposal, followed by correspondence and further meetings.

Groundwork for the International Network on Soil Fertility and Sustainable Rice Farming (Appendix 1), for example, was laid during the International Rice Research Conference held at the International Rice Research Institute at Los Baños, Philippines, in 1975. The African Research Network for Agricultural Byproducts (ARNAB), devoted to the better use of agricultural by-products for livestock feed, is an outgrowth of contacts made during the 1981 meeting of the Association for the Advancement of Agricultural Sciences in Africa (AAASA) in Douala, Cameroon. The Micro-Biological Resources Centers (MIRCENs) are an outgrowth of the 1973 United Nations

Conference on the Human Environment in Stockholm, and the Pasture Network for Eastern and Southern Africa (PANESA) traces its origins to a workshop on Pastures Research in Eastern and Southern Africa held in Harare, Zimbabwe, in 1984.

A congress organized by the Vienna-based International Union of Forestry Research Organizations (IUFRO) at Kyoto, Japan, in 1981 solidified concerns that forestry research had largely neglected the tropics and new strategies were needed for developing countries. A paper presented by the World Bank and the Food and Agriculture Organization of the United Nations at the congress underlined the importance of site-specific research in the Third World and the need to foster collaborative research.[1] A combination of grass-roots scientific interest and donor support led to the creation of the Special Program for Developing Countries (SPDC) in 1981. At the eighteenth world IUFRO congress in Yugoslavia in 1986, the network's name was changed to the International Council for Forestry Research and Extension (INCOFORE) and the cooperative effort was formalized.

Training courses can also provide the setting for starting a network. In 1977, a potato production training course in Toluca Valley, Mexico, provided the setting for discussions on ways to pool resources of national potato programs in Central America and the Caribbean (ISNAR, 1985). A year later, this encounter led to the creation of PRECODEPA, a multipurpose, collaborative research network that now connects scientists in ten countries.

Early Growth

As networks get under way, leadership, communications, program planning, active involvement of members, and delineating a growth and operational strategy are usually the first challenges faced by the coordinator or steering committee. To be successful, a network must establish what it wants to do and what it wants to be. Somehow, the network must find a way to provide for leadership, establish guiding values and principles, and develop a strategy for growth and development.

Leadership is one of the earliest problems to be faced. Donors or

1. "Forestry Research Needs in Developing Countries—Time for a Reappraisal?" paper presented at the seventeenth IUFRO (International Union of Forestry Research Organizations) Congress, Kyoto, Japan, 6–17 September 1981, World Bank, Washington, D.C./United Nations Food and Agriculture Organization, Rome.

international agricultural research centers are often heavily involved in the leadership of a network, at least in its launching and early growth. In many cases, donors and international agricultural research centers directly or indirectly lead networks because of their financial resources and experience with other networks.

Enthusiastic individuals whose vision for a network is tempered with realism are ideal network shepherds. Strong leaders are essential to help generate momentum for a network during its early growth. Resiliency also helps overcome problems that will inevitably arise and can retard progress unless handled with patience and persistence.

Ensuring rapid dissemination of research results and other information is one of the first tasks of a network leader. Often information flow is initiated through a newsletter, thus early development of a communications system is important in the nurturing of a network.

A network's early values and principles are often a blend derived from the network's instigator and the organizing committee. Some values and principles also result from early patterns in the network's development.

Each network in its early growth must find a way to respond creatively to its ever-changing environment. Identifying and then catering to major network stakeholders, especially clients and donors, should be at the top of the agenda from the outset. A properly drawn-up strategy helps a network find its footing. Planned approaches that seemed feasible are typically modified or dropped, new directions may be started, and publications such as training manuals or research protocols are usually circulated.

A growing network is a learning environment for participants. Visits by the coordinator to participants at their places of work can help stimulate the interest of individual scientists as well as solidify support of local administrators for network activities. Early identification of commonly shared problems justifies the network's research goals and helps mobilize support.

During the early years of a network, a great deal of its business is conducted at the coordinator's host institution. The coordinator and the associated policy committee interact frequently with network members as new projects are launched. A network needs a hands-on approach by the coordinator until operations become routine. Annual meetings, newsletters, and monitoring tours soon become part of the regular activities of all but information exchange networks. Also in the early stages of a network, plans may be laid for specialized training courses to bridge gaps in expertise.

As the network progresses and some members start to take on additional responsibilities, steps toward decentralization and a lightening of the coordinator's duties may be warranted. If transformation to a more integrated, self-regulating network is foreseen in the planning or founding documents, some moves to accomplish this, such as drawing up protocols, are usually under way.

Maturity

A mature network should be at its most productive stage. Most problems associated with articulation and differentiation should have been sorted out by this time. Articulation between network members should be well developed and communications should be direct and smooth. Persons who have received specialized training should be in place at their home institutions and well advanced in their research efforts (Figure 8). Participants should be enthusiastically engaged in network activities, motivated by the rewards of collaboration. Mature and productive networks usually enjoy the support and confidence of local administrators and external donors. Successful networks are an asset to the individual scientist and his or her workplace.

Mature networks may be at their prime, but they are also vulnerable. Complacency may begin to set in, especially as patterns of operation become routine. Instead of questioning procedures and seeking more efficient ways of conducting business, network participants may fail to keep abreast of new technologies and research techniques. Well-attended annual meetings at which project leaders report on their activities provide opportunities for critical analysis and joint planning, so vital to rejuvenate a network with fresh ideas and insights. A great danger to all networks is the strangulation caused by increasing layers of bureaucracy. Efficiency and streamlined procedures should be paramount in a network at all times, but especially at maturity.

In most mature networks, coordinators are able to relax their control without interrupting the smooth functioning of the network. Strong radial nodes and new linkages have often emerged, including subnetworks, that are capable of carrying on activities without constant supervision. Such developments are to be welcomed, but they nevertheless create the need for some adjustments and bring to the surface fresh management challenges. As some research tasks are accomplished, shifts in direction require careful planning and new assessments of resources and the scientific capabilities of participants. For networks that started as nonpeer relationships, some move toward

Figure 8. Collaborators in the International Network on Genetic Enhancement of Rice, a relatively mature network with well-trained participants, reviewing trials of cold-tolerant rice varieties in Palampur, India. Courtesy of the International Rice Research Institute.

transfer of leadership to recently strengthened national programs is usually under way as the network matures.

Dissolution

Most networks will eventually disband so that staff and other resources are free to regroup and confront emerging problems. Networks set up to tackle specific problems should dissolve once the task has been accomplished or the problem proves intractable. Some networks, especially those initiated to achieve a specific purpose, may reach their planned goals and then shift their work to other fields. In some cases, however, a network may be dissolved after the work is completed. Other networks, particularly germplasm evaluation nurseries, may need to be ongoing enterprises.

Most networks should consider themselves temporary organizations. A limited life span is particularly desirable for networks set up to investigate narrowly defined issues. When the task is clearly defined and tackled in a planned manner, the collaborative effort can be executed with milestone events in mind that can serve as midcourse evaluation points. Milestone events provide opportunities to make corrections or terminate the joint effort.

A major concern in determining the duration of a network will be the nature of the task or set of related problems to be addressed. Some networks address problems on a project rather than a program basis. We define a project as having well-defined research goals, with a clear beginning and an end. By contrast, a program is broader in scope, with wider boundaries and a more indefinite life span. Indeed, most research programs embrace several projects, each addressing parts of the research agenda. Networks along the lines of a program usually endure longer than collaborative efforts focusing on specific topics or projects.

Although some networks should probably have a limited life span, only a few networks in international agricultural research have thus far ceased operations. The IRRI-led International Agro-Economics Network was phased out after completing its work program. This network identified constraints to rice production in Asia and thereby helped lay the groundwork for the larger Asian Rice Farming Systems Network (Appendix 1). The International Committee on Land Clearing and Development in the Tropics (ICLCD), an information network instigated by Rattan Lal at IITA in 1982, was folded in 1985 with the establishment of the more comprehensive International Board for Soils Research and Management, headquartered in Bangkok, Thailand. Another network that ceased operation was the Nutrition Collaborative Research Support Program, which began in 1981 and was phased out by the U.S. Agency for International Development in 1988.

The apparent endurance of so many networks can be attributed partly to their relative youth. Most agricultural research networks operating internationally are less than fifteen years old (Appendix 1). But the tenacity of networks may also be due in part to a reluctance by members or donors to part with an enterprise, even if some of its stated goals are not being reached.

The endurance of networks may attest to their flexibility and vitality. Instead of dissolving a network, it may make more sense to transform the joint enterprise to address new issues. The following questions must be asked before a network is folded or transformed: Has the task been completed? What parts of the task remain undone? Do opportunities for fruitful collaboration still exist in some areas? What changes are necessary to meet the current situation? If the network is to be phased out, what concerns need to be addressed in the transition? How can the strengths and major products of the network be used most effectively during its dissolution or transformation?

Donors and the host institutions of participants often dictate the ultimate fate of a network because they hold the purse strings. The rapid proliferation of networks and their rising costs are bound to cause some donors eventually to take a hard look at some of the networks they sponsor and terminate support for those deemed less worthy. It behooves the membership of a network periodically to assess progress toward its goals and check how closely it follows principles for success.

5

Principles for Success

Networking generally appeals to scientists because they like to cooperate and share information with colleagues. The idea of dividing research tasks into manageable proportions, visiting study sites, and participating in regular workshops is attractive. But networking is also time-consuming, and collaborators are likely to abandon the effort if payoffs do not ensue. Scientists constantly assess the opportunity costs of their various endeavors; those projects that never bear fruit or cease to be productive are likely to be dropped. But the chances of a network foundering can be minimized if certain principles or guidelines are followed.

Although networking in agricultural research is a fairly recent development, valuable lessons can be extracted from the results of collaborative efforts worldwide. Scientists and organizations have examined agricultural research networks and have extracted principles for successful collaboration (Winkelmann, 1987). Over two dozen characteristics or attributes for viable networks have been identified (Table 4).

In Chapter 3, we summarized major tenets for effective management of networks. Here we explore fourteen main principles underlying the overall success of networks. Fourteen is not a magic number. As in plant and animal taxonomy, one finds "splitters" and "lumpers." It could be argued that there are more than twenty-five principles for successful networking, or that they can easily be condensed to half a dozen. Some of the fourteen principles explored here are interrelated and could conceivably be combined. Also, not all the principles apply

Table 4. Principles considered important for networking by eight evaluators*

Principle	Author/Organization							
	1	2	3	4	5	6	7	8
Strong self-interest	X	X	X	X	X	X	X	X
Effective coordinator	X	X	X	X	X	X	X	X
Clearly defined problem	X	X	X	X		X	X	X
Participants commit resources	X	X		X		X	X	X
External funding	X	X	X			X	X	X
Problem widely shared	X	X						X
Scientific capacity to contribute	X	X					X	X
Effective advisory group			X	X			X	
Scope for new ideas/free exchange			X	X				
Sufficient new ideas and technologies			X					X
Participants involved in planning and management		X						X
Regional scope			X					
Linkages to basic research			X					
Clear strategy as well as theme				X				
Training and monitoring				X			X	
Common constraints					X			
Capacity to adapt and diffuse					X			
Access to other networks					X			
Long horizons						X		
Extra funding for national programs			X					
Authoritative founding document								X
Realistic research agenda								X
Regular workshops and meetings								X
Relatively stable membership								X
Flexible research and management								X

* *Author/Organization*
1. Plucknett and Smith, 1984
2. Greenland et al., 1987
3. Zandstra, 1986
4. SPAAR, 1986
5. ISNAR, 1985
6. FAO, 1985
7. Valverde, 1988
8. Plucknett, Smith, and Ozgediz, this chapter

to all networks. Information exchange networks, for example, are often less complex and fewer principles apply to them. But in general, the more principles that are adhered to by a cooperative or collaborative research effort, the more likely the joint enterprise will be worthwhile.

To be successful, scientific consultation and collaborative research networks should incorporate fourteen main principles: (1) the problem is widely shared, (2) participants are motivated by self-interest, (3) participants are involved in planning and management of the network, (4) the problem or focus of the network is clearly defined, (5) a baseline study is undertaken to produce an authoritative founding document, (6) a realistic research agenda is drawn up, (7) research and management are flexible, (8) the network is constantly infused with new ideas and technologies, (9) regular workshops or conferences are held to provide opportunities for assessing progress and discussing problems, (10) collaborators contribute resources, (11) external funding is provided to facilitate travel, training, and meetings, (12) collaborators have sufficient training and expertise to contribute effectively, (13) the network's membership is relatively stable, and (14) leadership is efficient and enlightened. These principles take into account all of the components of the conceptual model discussed in Chapter 3 and underscore the importance of harmonizing all facets of a network's activities and resources.

An analysis of the ingredients of successful networking will be useful to those already involved in networks and others who plan to start or join one. Identifying and analyzing the underlying strands that make for a solid network will help collaborators assess the health of their networks and will guide policy makers when deciding whether to support existing networks or set up new ones.

Problem Widely Shared

At the top of the checklist in deciding whether to set up a network is to determine whether the problem or research subject is widely shared. The problem can be at the national, regional, or global level, and the scope of the network will be adjusted accordingly. Commonly shared problems have provided the "glue" for the Southeast Asian Universities Agroecosystem Network (SUAN), an independently planned network linking farming systems research in seven countries (Rambo and Sajise, 1985).

A widely shared problem is also likely to spark the interest of the donor community, whose support is vital for the development and maintenance of international networks. Highly localized phenomena are not appropriate for international networking because only one party would benefit from the joint research and external support is difficult to obtain. One of the reasons that donors from international agencies and foundations find networks appealing is that they identify issues of common concern and thus investments in the research effort are more likely to have a widespread impact. The supportive role of donors is also more likely to be noticed when they are involved in matters of widespread concern.

Self-Interest

An additional benefit of selecting a widely shared problem is that participants are more likely to be motivated by self-interest. Collaborative programs work best when the players have a vested interest in the outcome. Networks germinating from grass-roots efforts are more effective than those mandated from above. No one likes to do someone else's work without recompense. In successful networks, work and results are shared, and tangible payoffs stimulate further activity. In international nurseries, for example, collaborators plant other people's plant materials as well as their own. Sometimes promising foreign material is spotted in such trials and is introduced into local breeding programs. Some costs accrue to individuals when planting other scientists' materials, but occasionally the material is useful to more than one of the network participants.

The principle of self-interest also applies to the participants' place of employment. Collaboration will not proceed far if directors, deans, or ministers do not support the participants' desire to take part in a network. Participants will need permission to engage in collaborative activities, and their institutions will be called on to provide staff, space, or other resources for the network. It is important, therefore, for network leaders to approach management and administration at research institutions to learn about their interests and priorities and to what extent scientists involved in networking can further those institutional goals. Parallel motivational development of researchers and their bosses is essential. For example, external support for training and staff travel to professional meetings, one of the fourteen principles, is a major selling point for facilitating staff engagement in collaborative research networks.

We can differentiate between three kinds of self-interest. A participant can be motivated by an opportunity to advance his or her career. Such ambitions are by no means undesirable, or less worthy, motivating forces than, for example, a desire to advance knowledge. Personal concerns, however, can warp a network unless they are congruent with larger network goals (Winkelmann, 1987). A second type of self-interest is institutional, which we have already pointed out is essential if the participant is to be free to engage in a network. Permission and support from participants' host institutions are necessary so resources will be available for their collaboration in outside activities. Finally, participants are motivated by the satisfaction of gaining new knowledge and obtaining fresh insights from collaboration in a network.

The principle of self-interest applies to collaboration in other scientific fields. In the case of the successful international effort in West Africa to control onchocerciasis, a human disease that can lead to blindness, self-interest has been pinpointed as a major factor in the ability to eradicate the vectors of the disease from many areas (Walsh, 1986a). The ongoing onchocerciasis control project is opening up large areas of land for farming in a region suffering from chronic food deficits.

Participants Involved in Management

One way to ensure that network participants are highly motivated is to involve them early in planning and managing the collaborative effort. When network members are invited to help establish priorities, the joint enterprise is more likely to be relevant to the needs of clients—the participants and, ultimately, farmers and consumers of agricultural products. Also, when network participants are involved in managing the joint exercise they are more likely to feel that the network is theirs, rather than handed to them as a fait accompli (Greenland et al., 1987).

Involvement of participants in management has helped sustain numerous agricultural research networks. Five regional potato networks, fostered by CIP, have policy and technical advisory committees composed of representatives from each of the participating countries. Networks coordinated by IRRI also have advisory committees or working groups that invite management inputs from participating countries (Greenland et al., 1987). Self-governance is preferable to a constant hands-on approach by a donor or lead institution. The more mature a network is, the more smoothly self-governance works, but it should be encouraged from the start. Administrative and other costs are reduced when the network handles its own affairs well and does not need outside trouble-shooters.

Clear Definition of the Problem

The problem or scope of the network's task must be clearly defined at the outset or the collaborative effort is likely to become sidetracked and diffuse. The initial flush of enthusiasm for cooperation can blind participants to the need to delineate carefully the task at hand. If the problem is not properly identified, the operation may turn out to be unmanageable and much time will be wasted. When the "hare" that everyone can chase enthusiastically has been spotted, participants are more likely to pull together as a team.

Founding Document

A baseline study that explores the scope of a problem and identifies key participants is essential for scientific consultation and collaborative research networks. An objective survey before a network is formed helps define the research problem and serves as the foundation for drawing up a realistic work agenda. An extensive study of livestock trypanosomiasis in Africa commissioned by FAO and UNEP, for example, laid much of the groundwork for the highly productive Trypantolerant Livestock Network coordinated by the International Livestock Center for Africa (ILCA/FAO/UNEP, 1979).

A founding document may take various forms, from a scholarly book to a mimeographed report, but it must constitute an objective analysis of the problem. A founding document is not a network proposal. A credible proposal emerges from a thorough baseline survey that summarizes the state of knowledge of a particular subject or problem. To reduce chances for failure, donors interested in supporting a new network might first provide seed money to conduct an initial diagnostic survey so that the network proponents have a firmer grasp of the problem they are attempting to address. Furthermore, such a study would give donors a better understanding of whether the network's objectives are feasible.

Realistic Research Agenda

Every network needs a realistic research agenda. Strategic planning is not high on the list of enjoyable tasks for scientists, but some commonly agreed-upon work plan is essential if the network's efforts are to proceed coherently. A strategic planning exercise helps participants assess their collective strengths and weaknesses and provides guideposts for research and management activities.

The task of drawing up a workable strategic plan is facilitated when the collaborative effort is correctly focused. Typically, the problem or task is divided up among participants according to their particular interests, expertise, and facilities. Each collaborator is thus responsible for pursuing one component or piece of the puzzle that is commensurate with his or her abilities.

Flexibility

Flexibility should be built into the research agenda and management of a network so that the collaborative effort responds to the ever-shifting contours of research frontiers and farming environments. Fortunately, this principle is widely adhered to in agricultural research networks and is a major reason for their popularity (Oram, 1980). Regular self-examinations and periodic midcourse corrections are hallmarks of productive networks. Flexibility and the ability to be self-critical go hand in hand.

The International Network on Genetic Enhancement of Rice (INGER), as is typical for international material exchange networks involving crop germplasm, adds nurseries when sufficient requests are made and drops them when interest wanes. In any given year, INGER operates from twelve to twenty-eight nurseries. In response to growing interest, INGER added the International Rainfed Rice Shallow Water Observational Nursery (IRRSWON) and the International Rainfed Rice Shallow Water Yield Nursery (IRRSWYN) in 1978 and 1981, respectively. INGER started a gall midge nursery in 1975 but dropped it in 1979 because collaborators did not enter enough promising material; however, the widespread pest was serious enough to warrant starting the nursery again in 1987. INGER halted a nursery focused on rice adapted to high temperatures and salinity because only Egypt and Pakistan showed strong interest in it.

Another example of the flexibility of networks is PRECODEPA's decision to include potato processing and to admit Cuba and El Salvador in 1983 and Haiti in 1985 (CIP, 1985a, 1986). PRECODEPA, started in 1978, has remained relevant to the needs of potato research in Central America and the Caribbean because of its adaptability. This flexibility is attributed to its simple operational structure, a clear understanding of the region's needs, an ability to allocate and manage funds, periodic project evaluations, and constant dialogue and interaction among participants (CIP, 1985a:35).

Networks could have a tendency to lock into obsolete methodologies or problems and become closed clubs. But this rarely happens.

Coordinators, working in conjunction with their advisory bodies, are generally responsive to new requests to join the network and to open new avenues for research. And various mechanisms are in place to receive and act upon feedback from participants. SUAN, for example, is reaching out to establish links with scientists involved in agroecological research in East and South Asia (Rambo and Sajise, 1985). As a first step, information will be exchanged with other groups, but other countries may join the network. SUAN would like eventually to forge ties with scientists studying rural ecology in tropical parts of Africa and Latin America.

New Ideas

It may seem obvious that any worthwhile scientific research effort will be constantly infused with fresh ideas, insights, and, in some cases, technologies. As enthusiasm for networking builds, the perception may arise that they can be a cure-all for scientific malaise. Although networks have some promising properties not found in individual research efforts, they are not immune to stagnation or a blindered approach.

To remain vital and relevant, collaborative research needs to be stimulated by challenging concepts and the development of new methodologies and technologies (Greenland et al., 1987). Strong national programs in a network have a major responsibility to display leadership in generating new ideas and introducing the latest techniques for field and laboratory work. Flexibility, a desirable attribute already discussed, can help ensure an openness to novel viewpoints and research procedures.

Another way to encourage a generous flow of new ideas and techniques is to forge upstream linkages with basic research institutions, an attribute considered important for networks by at least one authority (Table 4). CIP and CIMMYT regularly contract out basic research to institutions better equipped to do advanced projects such as genetic engineering. Networks coordinated by international centers are thus able to benefit indirectly from these upstream linkages, but there is also scope for similar relationships by networks, particularly in the Third World. Networks composed of participants from developing countries would do well to establish mutually beneficial collaboration with research centers in industrial countries or advanced scientific institutions in such developing nations as Brazil, India, and Mexico.

Regular Meetings

Frequent face-to-face meetings, not just telephone or mail communication, are essential to further research progress in networks. One of the most enduring values of networks is the opportunity, and rationale, for regular meetings with scientists to discuss common problems (Maddox, 1987). Workshops and regular meetings of network participants foster the exchange of new ideas and techniques. Too frequent meetings, however, are counterproductive because they drain resources and take researchers' valuable time. Constant personal contact is particularly important in collaborative research networks since departures from the jointly planned agenda can lead to dissipation of the enterprise. Midcourse corrections to the research effort can be more easily agreed to and administered when regular meetings are held.

Two basic kinds of meetings are essential for the smooth functioning of material exchange, scientific consultation, and collaborative research networks. Steering committees meet at least annually to review programs and adjust overall policy. Scientists also need to get together once or twice a year to present results from their experiments and research and to learn about other participants' solutions to problems. Conferences or workshops are appropriate mechanisms for promoting interactions between network collaborators.

Frequent regional workshops organized by SUAN have been a key factor in the growth and enthusiasm characteristic of the network (Rambo and Sajise, 1985). SUAN's scientific meetings help break down cultural barriers between individuals and foster respect and improved communications between the various disciplines involved in farming systems and resource management research.

Frequent meetings are particularly important for networks with scientists from a broad range of disciplines. Misunderstandings and concerns between disciplines can often be cleared up in round-table discussions. Frequent informal meetings are also especially helpful in the early stages of a network when unexpected difficulties often arise. Once a collaborative effort is well under way, meetings can be scheduled at longer intervals.

Collaborators Contribute Resources

One criterion used in deciding whether a candidate can join the network is whether the person or institution will voluntarily contribute resources to network activities. Indeed, one measure of the collaborators' interest in a network is their willingness to provide facilities, staff, and other resources. A desire to cooperate should be backed up by tangible contributions, or the network becomes a one-way flow of resources and information. This principle is followed by business corporations or partnerships; each partner must contribute a fair share of resources for the association to remain cohesive.

Most agricultural research networks operate on the premise that collaborators will contribute a fair share of staff, facilities, and other investments. When the principles of selecting a widely shared problem and self-interest are incorporated in network planning, there is usually little difficulty in eliciting offers for the use of land, buildings, equipment, and personnel.

National programs in the Third World are often hampered by poor facilities and operating budgets. But pleading poverty is no excuse for failing to do one's fair share of the work. Budget directors and scientific directors should give networks high priority, or withdraw from them. The West African Farming Systems Research Network (Appendix 1) got off to a slow start in part because of a lack of fuel for vehicles for researchers to conduct surveys. The seventeen-nation WAFSRN began in 1982 but did not hold its first workshop until 1986. Much baseline data for the network awaits collection and analysis.

Collaborators should not be bought. A guiding philosophy behind the international maize nurseries of CIMMYT is that neither individuals nor organizations are ever paid to plant CIMMYT materials. This principle was firmly laid down by Ernest Sprague, former director of CIMMYT's maize program, and is still adhered to in regions served by the international center (M. Collinson, pers. comm.). Curiosity and the potential of finding rewarding germplasm motivate subscribers to CIMMYT's nurseries. Motivation is not the only question if participants are paid to join a network. Funding requests and how the money is spent would be subject to query, and such inquiries would surely dampen any congenial spirit.

Although contribution of resources by participants in collaborative research is the acid test for networks, partial exceptions to the principle do occur. At some sites in soil fertility networks, for example, participants are provided with fertilizer and seeds to run trials. Furthermore, sometimes technicians may be paid by the network coordinator to conduct the experiments. In the case of IRRI's Industrial Extension Network (IIEN), manufacturers of farm equipment in Indonesia have been partly subsidized to produce machinery prototypes for testing by farmers (Reddy, 1984). Manufacturers of equipment, as well as engineers involved in designing and testing equipment, are included in IIEN. And in the Philippines, several companies in the Manila area have produced various farm implements and machines for distribution by the national extension service. A guaranteed order by the federal government for the machinery acts as a subsidy.

Relatively weak national agricultural research systems may need some support for local costs of networking. For example, USAID missions in Burundi, Rwanda, and Uganda underwrite some of the costs of local scientists' participation in a potato research network for the Great Lakes Region of Africa (PRAPAC). The principle of contributions by members still applies, however, because salaries of national program scientists involved in PRAPAC are covered by their respective governments and national experiment stations are used as research sites.

Prolonged and heavy subsidies, though, are not a good formula for networking. Collaborators should be weaned from direct payments or substantial transference of equipment and supplies as soon as possible, or they may be participating in networks for the wrong reasons. Several manufacturers of farm equipment in Metro Manila, the Philippines, for example, stopped making certain farming implements when no new orders were received from the government.

External Funding

Although institutions participating in a network should bear a sizable portion of the cost of operating the cooperative effort, some outside financial assistance is usually required to ignite and maintain a collaborative effort (CIP, 1984b, 1985a). The proportion of funds provided by external sources varies between networks and at different stages of a network. Information networks generally do not require much, or any, outside support to sustain the compilation and mailing of newsletters. Collaborative research networks, however, often require fairly sizable contributions from external donors. Funds are needed for coordination, travel, research planning meetings, steering committee meetings, and to cover at least part of the research costs for participants. It is not uncommon for outside donors such as United Nations organizations or bilateral aid agencies to supply more than half the operating costs of an international network (Table 5). For example, external funding accounted for 60 percent of the $2 million spent by PRECODEPA during 1980–84 (CIP, 1985a).

Table 5. Proportion of outside funding for some international agricultural research networks

Network	Budget ($ million)	Period	External funds (%)	Donor
PRECODEPA	2.00	1980–84	60	SDC
RNAM	.80	1986	70	UNDP (50%); Japan and Australia (20%)
SAPPRAD	2.30	1982–86	59	ADAB, Australia (54%); CIP (5%)
CRSP—Aquaculture	1.45	1983–84	50	USAID
CRSP—Bean/Cowpea	4.54	1983–84	65	USAID
CRSP—Peanut	2.2	1989	73	USAID
CRSP—Pond Dynamics	1.06	1988	88	USAID
CRSP—Small Ruminants	5.58	1983–84	51	USAID
CRSP—Soil Management	2.71	1983–84	69	USAID
CRSP—Sorghum/Millet	4.09	1983–84	66	USAID

Note: For network acronyms see Appendix 1. Donor acronyms: ADAB = Australian Development Assistance Bureau, CIP = Centro Internacional de la Papa, SDC = Swiss Development Cooperation, UNDP = United Nations Development Program, USAID = U.S. Agency for International Development.

Seed money to start an international network in the Third World often comes from external donors. A workshop is commonly used to gauge interest in starting up a network. Even a modest meeting with a few dozen individuals can easily cost close to $100,000. In 1983, it cost approximately $70,000 to transport and lodge thirty West Africans to participate in a workshop to discuss the idea of setting up a farming systems network at the International Institute of Tropical Agriculture in Ibadan, Nigeria.

Outside assistance is also crucial to maintain international networks in the Third World. Citizens of developing countries frequently encounter bureaucratic barriers to obtaining passports, exit visas, and foreign exchange for travel abroad. When the trip is to be paid for by another party, however, the process of securing the necessary documents and foreign currency is usually expedited. External funds are also sometimes needed to import equipment and supplies not available locally, such as for participants in PRAPAC. Donors in industrial countries thus play a key role in providing networks with funds for travel, communication, and international shipment of technology.

One of the benefits of networks is that they can be a cost-effective use of research funds because existing facilities and staff are relied on heavily to carry out the work. Nevertheless, networks can be expensive to operate. Costs soar the larger the network, particularly if international travel and equipment purchases are involved, hence the dependence on external support (Table 6). The U.S. Agency for International Development, the Canadian International Development

Table 6. External funding and funding sources for some international agricultural research networks

Network	External funds ($ million)	Period	Main source
AFRENA	4.5	1988–93	World Bank, USAID, IDRC, CIDA, SAREC, IFAD
CIMMYT/ESA	0.25	1988	USAID, CIDA
CRSP—Bean/Cowpea	2.6	1989	USAID
CRSP—Fisheries Stock Assessment	0.7	1989	USAID
CRSP—Peanut	2.2	1989	USAID
CRSP—Pond Dynamics	0.92	1989	USAID
CRSP—Small Ruminants	2.8	1989	USAID
CRSP—Soil Management	2.1	1989	USAID
CRSP—Sorghum/Millet	2.7	1989	USAID
IBSNAT	5.4	1987–92	USAID, UH
INIBAP	1.5	1989	France, Belgium, IDRC
INSURF	0.34	1988	SDC
INGER	8.79	1985–89	UNDP
IWBP	0.7	1985–90	IOCCC
NifTAL	8.73	1982–86	USAID, NSF
PRECODEPA	1.92	1978–83	SDC
RISPAL	0.24	1988	IDRC, IICA, CATIE, INIAA
RNAM	3.03	1987–91	UNDP, Australia, Japan
Trypanotolerant Livestock Network	0.35	1987	AfDB, EEC, ODA, GTZ, Belgium, SDC, France, Netherlands

Note: For network acronyms see Appendix 1. Donor acronyms: AfDB = African Development Bank, CATIE = Centro Agronomico Tropical de Investigación y Enseñanza, CIDA = Canadian International Development Agency, EEC = European Economic Community, GTZ = Gesellschaft für Technische Zusammenarbeit, IDRC = International Development Research Centre, IFAD = International Fund for Agricultural Development, IICA = Instituto Interamericano de Cooperación para la Agricultura, INIAA = Instituto Nacional de Investigaciónes Agropecuaria y Agroforestal, IOCCC = International Office of Cocoa, Chocolate, and Confectionary Sugar, NSF = National Science Foundation, ODA = Overseas Development Administration, SAREC = Swedish Agency for Research Cooperation with Developing Countries, SDC = Swiss Development Cooperation, UH = University of Hawaii, UNDP = United Nations Development Program, USAID = U.S. Agency for International Development.

Agency, and CIMMYT provide $250,000 a year to the fourteen-nation CIMMYT Eastern and Southern Africa Economics Program (CEAREP). The Trypanotolerant Livestock Network, a ten-country collaborative effort coordinated by ILCA, receives $350,000 per annum from a consortium of donors, mostly from Europe.[1] Although the Trypanotolerant Livestock Network covers fewer countries than CEAREP, it is more costly because it uses laboratory equipment and supplies. The much larger INGER network, coordinated by IRRI, uses about $7.7 million from the U.N. Development Program every five years.

Adequate Training

Another cardinal rule of networking is that participants should be sufficiently prepared for the task. Linking a group of poorly trained people into a consortium effort can be counterproductive. Not all participants in a collaborative effort are likely to be of the same caliber, but they should have sufficient training and facilities to make a contribution. If the capacity of participants varies markedly, the output of the network is likely to be at the level of the lowest common denominator. In any multiple-step process, typical of activities in organizations such as networks, educational, and scientific establishments, the operation may be only as efficient as its least efficient step (Vaz, 1987). A net is only as strong as its weakest knot (Valverde, 1988). But the weakest knot concept is not true in all cases. In a well-run network the network itself may help compensate for, or reinforce, the weakest knot.

If a weak linkage cannot be immediately corrected, the network should be rewired so that critical processes can continue with minimum hindrance. Multipurpose commodity networks sometimes shift the responsibility of an assigned task to another country if a member institution is encountering difficulties in fulfilling its part of the research effort.

1. External donors to the Trypanotolerant Livestock Network include the African Development Bank (for work in the Gambia and Senegal), the European Economic Community (EEC), Britain's Overseas Development Administration (ODA), the German Agency for Technical Cooperation (GTZ—Gesellschaft für Technische Zusammenarbeit), Belgium's Administration Generale de la Cooperation au Developpement, the governments of the Netherlands (for research in Zambia) and Switzerland, and May and Baker Ltd. of the United Kingdom.

"Stars" in a network can pull the communal effort only so far. Networks are thus no substitute for the long-term upgrading of scientific capability, particularly in developing countries. Network courses can help bridge training gaps, but better instruction in schools and universities is still needed to help people gain a sense of the broader picture and obtain the necessary skills to pursue problems in depth. In the Third World, agricultural research networks are best developed in Asia and Latin America because national programs are generally stronger in those regions.

The scientific capabilities of network members vary considerably. This is a major reason why so many networks operate short training courses or facilitate university-level training for some members. CIP takes a liberal attitude by welcoming weak members to the potato research networks (Ezeta, n.d.). Weak national research systems need to acquire technologies and can serve as useful testing sites for germplasm or machinery. A donor might only want to support a network that was composed exclusively of strong research institutions or individuals, but in almost all cases networks also serve to upgrade the capacities of members.

Stable Membership

Commitment can be as important as technical caliber to a collaborative research effort. Stable membership helps promote continuity and a collegial atmosphere, so important for the ultimate success of a network. In long-term networks such as international nurseries, turnover of participants is not a crucial issue, but for short-term networks set up to tackle specific problems, rapid changeover in the ranks of collaborators can stall the research effort. The relative stability of personnel in potato research in Central America and the Caribbean is one reason for the success and mutual satisfaction of participants in PRECODEPA (ISNAR, 1985).

Scientists have occasionally expressed dismay at the loss of a valued colleague in a collaborative project through circumstances beyond the network's control. In developing countries especially, the brightest scientists are often promoted rapidly to leadership positions involving considerable administrative responsibilities. Burdened by increased paperwork and meetings, such scientists often withdraw from networks.

Promotion can further a scientist's career but takes sorely needed expertise out of the front lines of research. Subordinates may be assigned to take the places of those who leave, but they usually lack experience and familiarity with the collaborative research program. When subordinates or new members come on board, valuable time is lost as they attempt to pick up where their predecessors left off. In some networks, scientists have expressed irritation at the continuing need to educate new members about programs and procedures at committee meetings and workshops.

Strong Leadership

Finally, networks need to be guided by strong and sensitive leaders in whom the participants have confidence. Cooperation will wane if researchers feel that the leader is coercing them into a methodological straitjacket or if they do not receive recognition for their contributions. Such dissatisfaction is less likely when participants elect the network coordinator for a specified period. Coordinators elected by participants are more likely to have leadership skills than those appointed by outsiders, such as a donor. Initially, it may be necessary for a donor or coordinating institution to appoint a leader, but once the network is operational, coordinators are best elected by the members.

International centers often steer networks through their early development before ceding the helm to a national program. CIP, for example, provided a coordinator for PRECODEPA's first two years; thereafter, the advisory committee appointed a leader from a national program (ISNAR, 1985).

If leadership changes hands too frequently, however, research drives can stall and the network's cohesion suffers. When the research capability of institutes varies markedly, collaboration is best served by leaving the reins in the hands of the strongest participant. Large-scale international nurseries, for example, are usually coordinated by international centers because they are well endowed with staff and facilities to handle the large volume of seed shipments. And ILCA is still the coordinator for the African Research Network for Agricultural By-products after more than six years because the center owns printing facilities at its Addis Ababa headquarters and can easily publish in English and French.

A strong and enlightened leader is particularly important during the early years of a network (CIP, 1984b) or if the collaborative effort is taking on a new research problem. As the network matures and gains confidence, leadership may change hands without disrupting momentum. One advantage of rotating the leadership post is that it identifies and builds leaders in national programs. Furthermore, programmed leadership changes help dowse concerns about paternalism and authoritarianism.

In summary, an examination of principles for success in networking can be useful in network design and implementation, but not all of the principles will apply to every network.

6

Information Exchange Networks

A diverse range of networks has sprung up to address needs in various areas of agricultural research. Rather than attempt to enumerate exhaustively all known networks that are, or have been, operating in agricultural research, we sample a few networks in some depth. To give the reader some appreciation of the pervasive nature of cooperation and collaboration in agriculture, we select networks involved in a wide range of topics. Here we focus on information exchange networks.

For each network studied we describe how the components fit together to produce a cohesive organization. We are particularly interested in analyzing how the program is managed and what products are resulting and their impact. We make a distinction between performance and impact. A network may be performing well, with management working smoothly and the different components operating in unison, but no products may yet be available for farmers. The potential to deliver is the critical question here, particularly if the network is still young. Impact is a measure of the adoption of technologies by farmers or the institutions that serve them. Output alone is no measure of a network's success.

Information Exchange

Information exchange networks are essentially lists of participants with a coordinator responsible for gathering and collating relevant information for distribution and for updating mailing lists. Because information flows mostly to and from the coordinator and participating nodes, structure and management are relatively simple. Information exchange networks in international agricultural research tend to be highly specialized and relatively small. Some may include professional meetings such as PCCMCA at which scientists share information and discuss research progress.

Information exchange networks span the spectrum of agricultural topics from commodities to improved use of crop by-products (Table 7). Information exchange networks are easy to assemble and involve minimal costs so they sprout quickly. And with the advent of computerized mailing lists and newsletters, they are flourishing.

The impact of information exchange networks is hardest to gauge of all the networks because only information is disseminated and no coordinated research plans are established. Information exchange networks nevertheless provide a valuable mechanism for keeping people abreast of the latest developments in their fields. In some cases, they complement the work of material exchange, scientific consultation, and collaborative research networks, which have information dissemination programs of their own.

Information exchange networks are open-ended in that names can be added more or less at will, sometimes at no cost. Instead of dozens or hundreds of participants typical of most material exchange, scientific consultation, or collaborative research networks, information exchange networks can reach out to thousands of individuals. The flow of information is largely confined to a two-way stream between the coordinating hub and the nodes; interaction among the network

Table 7. Some information exchange networks dealing with agricultural research issues on an international scale

Network	Number of participating countries	Region
Animal Traction Research Network	25	West, East, southern Africa
FADINAP (Fertilizer Advisory Development and Information Network for Asia and the Pacific)	24	Asia, Pacific
Sago Advancement Group Office	13	Europe, Asia, Australia
SDI—ICRISAT	37	Semiarid tropics
SDI—ILCA	32	Africa

Note: See Appendix 1 for coordinators and starting dates of the networks.

participants is usually minimal. Information exchange networks are thus relatively passive, similar in design to a wheel without the rim.

The Sago Advancement Group Office started out as the Sago Palm Research Network in 1985 with the ultimate objective of establishing a regional institute to further research on the starchy palm (*Metroxylon sagu*). Sago palm is a significant food source for rural peoples in Malaysia, New Guinea, and some Pacific Islands. The net yield of calories from preparing sago palm starch is more than for most other starch-producing plants, and the towering palm also supplies useful thatch for houses. The pith from the trunk of the versatile palm can supply as much as 400 kilograms of starchy dough. Because sago palm fulfills a variety of nutritional and shelter needs, interest is growing in expanding its use.

The immediate aim of the Sago Advancement Group Office is to serve as a mechanism for disseminating information between groups interested in establishing sago palm as a major crop of the future. Since very few organizations are currently working on sago palm, the idea of following up with a scientific consultation or collaborative research network is not considered viable by some. Rather, a new center is considered necessary, possibly located in Papua New Guinea, with partial support from the Sago Palm Research Fund in Japan. An information exchange network devoted to sago palm is seen as a means of building a knowledge base to help launch a new research center.

Unlike the previous example, the Selective Dissemination of Information (SDI) service provided by ILCA is an information exchange network that has been set up solely to collate and pass on information (Table 7). SDI's purpose is to alert scientists in Africa about new publications in their fields of research. The research interests of each scientist in the network are profiled so that participants receive a monthly computer printout containing citations and abstracts of relevant journals and books (ILCA, 1981). The tailored printouts are generated from a data base composed of listings supplied by the Commonwealth Agricultural Bureaux International (CABI) and FAO's International Information Service on the Agricultural Sciences and Technology (AGRIS).

When ILCA's SDI service began in 1983, 250 people were enrolled. By 1986, the number of participants had grown to 550 and by 1988 there were 750 subscribers. Demand for ILCA's SDI service is expected to double or triple within the next few years, particularly as Francophone Africa becomes more familiar with this information outreach effort (Hardin et al., 1986).

The International Crops Research Institute for the Semi-Arid Tropics, located near Hyderabad, India, operates a similar information dissemination service for some of its mandate crops. The Selective Dissemination of Information service of ICRISAT now reaches thirty-seven countries in the semi-arid tropics. ICRISAT's Sorghum and Millets Information Center (SMIC) newsletter is sent in English to twelve hundred researchers and in French to three hundred scientists (ICRISAT, 1982).

Public Awareness Association for International Agricultural Research

An unusual network was launched in 1988 by information officers at international agricultural research centers, both within and outside the Consultative Group on International Agricultural Research. Dubbed the Public Awareness Association for International Agricultural Research, this network is designed to include a broad community of information officers at international centers, national programs, and donor agencies. The association was set up to further understanding about activities in the international agricultural research system. Target audiences of this network are the general public, environmental groups, donors, and policy makers.

The main goals of the new network, which held its first general meeting at CIMMYT in Mexico in 1988, are to develop new constituencies for international and national agricultural research and to counteract some of the misunderstanding about technologies developed for Third World farmers. The network is seen as a powerful tool to promote the cause of research. It is particularly timely in view of the pressure exerted by some groups in food-exporting countries to cut back on international assistance for agriculture because it is perceived as undercutting the potential for increased sales.

Membership in the Public Awareness Association is open to research and development organizations concerned with the global agricultural research system. At the inaugural meeting in Mexico, thirty-four representatives from thirteen international centers and seven donor agencies attended. The network is assisted in its work by the Public Awareness Council, a steering committee composed of members from five donors (Canada, Italy, France, the Rockefeller Foundation, and USAID) and three international centers (CIAT, CIP, and CIMMYT).

At the inaugural meeting of the association four main topics of particular interest and pertinence for information officers were identified: biotechnology, genetic resources, sustainability, and women in agriculture. For each topic, major issues and controversies were outlined, common interests among institutions were highlighted, and groups and institutions that should be reached with information about the activities of international centers in these areas were identified.

The association is an activist network with various projects, unlike most information exchange networks, which tend to be passive operations limited to exchanging ideas and research results. For this and other reasons, the association could easily be classified as a consultation or collaborative network, but we have included it in this chapter because the main focus is on information. The association has successfully garnered $1 million from the Italian government to raise awareness of crop genetic resource issues in Latin America. In 1990, a workshop for Latin American journalists will be held in Costa Rica to provide information on various aspects of conserving crop genetic diversity. Other projects include a seminar and travel fellowship program for journalists and science writers in North America, established with the assistance of the Rockefeller Foundation, and a public awareness office in Rome to serve Europe, which will be administered by the International Service for National Agricultural Research (ISNAR) and the International Board for Plant Genetic Resources (IBPGR).

Like most networks, emphasis is placed on effective use of existing funds to accommodate the association's goals. Funds for some of the network's specific projects have been obtained, but for the most part a realigning of priorities is envisaged using existing staff and facilities at international centers and other institutions. The association is seen as a cost-effective way of improving public outreach efforts of agricultural research centers and their supporters.

Specific proposals that came out of the inaugural meeting include the development of multimedia information packets on programs and accomplishments in the four priority topics, the compilation of an international data base of media contacts with interests in agricultural research and development, and the establishment of a data base on publications of special interest groups to pass on relevant information.

The Public Awareness Association has made impressive strides for a young network. Much of the credit for the successful launching of this multifaceted network rests with the dynamism and leadership of the founders of the association.

Outlook

More information exchange networks serving the international agricultural research system are likely to emerge. For example, efforts are under way to launch the African Agricultural Informational Resources Network (AAIRNET) to upgrade and better use information capabilities of national programs (Hailu, 1989). Specific objectives of AAIRNET include identifying and reaching end users for particular information services, assisting national programs in identifying funding resources to develop and maintain their information and documentation services, organizing short-term training courses, and publishing a newsletter.

The nature and scope of information exchange networks are likely to change dramatically as new information processing technologies become available. Instead of newsletters, electronic mail systems will increasingly tap into data bases. To speed up dissemination of results and reduce costs, some journals may be issued as magnetic codes that can be called up by computer software rather than the printed page. Some relatively inexpensive laptop computers rival the power and speed of yesterday's desktop models and surpass the capabilities of many mainframe computers of the early 1970s.

7

Material Exchange Networks

Material exchange networks are organizationally more complex than simple information exchange networks because they often have advisory bodies, organize monitoring tours, and are involved in training. Two main kinds of material exchange networks are discussed here: international nurseries that exchange and test plant germplasm and exchange networks that try out prototypes of farm machines and implements.

We do not wish to imply that material exchange networks are devoid of research since the testing of new machine designs or the screening of crop lines for resistance to diseases and pests involve experiments and verification. Indeed, a case can be made that some international nurseries such as INGER qualify as collaborative research networks because participants use a common methodology and jointly plan trials with the coordinator. We categorize them as material exchange networks because they are primarily concerned with the exchange and screening of germplasm, and most of the research is conducted after the material leaves the network and enters local breeding programs. Several regional crop improvement programs, such as those organized by CIMMYT, also employ international nurseries as part of an overall program to improve crop production.

International Nurseries

International nurseries have been widely developed to gauge plant germplasm for desirable characteristics such as high yield, resistance to pests, and tolerance to adverse soils and climate (Plucknett and Smith, 1986a). Dozens of major international nurseries serve the major food crops (Table 8). If one takes into account specialized nurseries subsumed under broader nursery programs for all crops undergoing intensive breeding, over a hundred international nurseries were probably operating in the late 1980s.

Crop breeders use international nurseries as recruiting grounds for

Table 8. Some international nurseries, their coordinators, the number of participating countries, and the regional coverage

Network	Crop	Coordinator	Number of countries	Region
AGLN	Pigeonpea, groundnut, chickpea	ICRISAT	10	Asia
ARSHAT	Sorghum	ICRISAT	3	India, Pakistan, Thailand
Confectionary Groundnut Varietal Trial	Groundnut	ICRISAT	14	World
EACSSN	Sorghum	ICRISAT	8	East Africa
EARSAM	Sorghum, millet	ICRISAT	8	East Africa
Groundnut Early Maturing Varietal Trial	Groundnut	ICRISAT	5	Asia, Africa
IBYAN	*Phaseolus* beans	CIAT	30	World
ICAT	Chickpea	ICRISAT	3	Korea, Colombia, Cape Verde
INGER	Rice	IRRI	80	World
International Maize Improvement Network	Maize	CIMMYT	90	World
International Wheat Nursery System	Wheat	CIMMYT	115	World
IPMAT	Pearl millet	ICRISAT	8	South Asia, Africa
IPMDRTP	Pearl millet	ICRISAT	4	South Asia, Africa
ISVAT	Sorghum	ICRISAT	37	World
PIN	Pigeonpea	ICRISAT	22	World

Note: A nursery is often planted at more than one site in a participating country. See Appendix 1 for network acronyms and starting dates and Appendix 2 for locations of network coordinators.

promising material for their various programs. Lines that perform well in nurseries may be released for farm use directly in several locations or used in further crossing (Chart 4). Modern farms rely on a steady stream of new cultivars to maintain productivity levels, and international nurseries are a crucial link in the multistep process of developing varieties (Plucknett and Smith, 1986b). International nurseries thus help in the global effort to sustain agricultural yields.

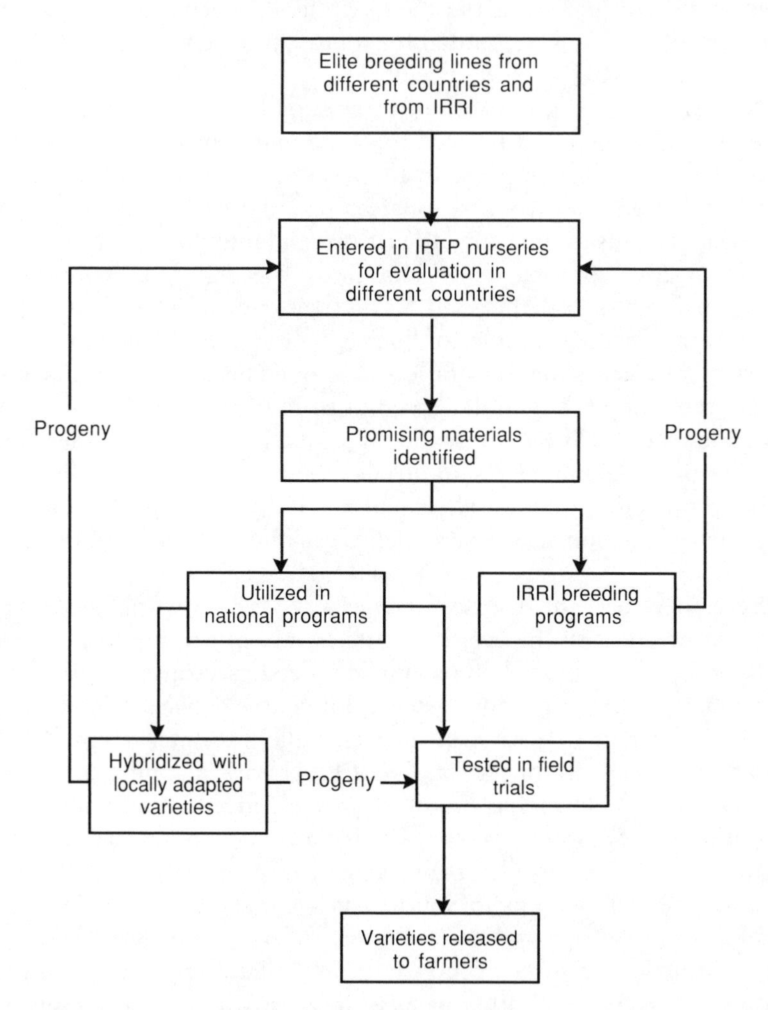

Chart 4. Flow chart for rice materials in the International Network on Genetic Enhancement of Rice, formerly the International Rice Testing Program, a material exchange network coordinated by the International Rice Research Institute, Los Baños, Philippines (information from Khush, 1984; Seshu, 1986)

Although nurseries have become increasingly important in the last few decades, they are not the only mechanism for sharing crop germplasm. IRRI, the world's largest "wholesaler" of rice breeding lines, has exchanged germplasm informally with national programs since the institute began in 1960. After IRRI started the International Rice Testing Program (IRTP) in 1975 (IRTP's name was changed to INGER in 1989), this international nursery network accounted for the bulk of seed samples dispatched from the institute. Between 1975 and 1983, for example, INGER accounted for 67 to 85 percent of seed samples shipped from IRRI.

Nurseries can be grouped into three broad categories. Yield nurseries are composed of advanced breeding materials that are near the end of the breeding process. In the case of INGER, yield nurseries concentrate on lines from national or regional trials that have been consistently outstanding for at least three planting seasons. Because yield nurseries concentrate on elite lines, breeders essentially select items of interest "off the shelf." Observational nurseries contain a more heterogeneous mixture of lines with a broader range of characteristics; breeders shop here for less developed lines that contain traits of interest to them. Yield data are not so important in observational nurseries. In INGER observational nurseries, plots are therefore smaller than yield nurseries and trials are not replicated. Finally, specialized nurseries are generally smaller and are often set up to tackle specific production challenges such as acidic soils or certain pests (Figure 9).

Specialized pest and disease nurseries serve as early warning posts by alerting breeders to the ever-changing nature of arthropod pest and pathogen populations. They also inform breeders, crop entomologists, and plant pathologists about genetic differences between populations of the same pest or pathogen. For example, the International Rice Brown Planthopper Nursery, an INGER network spanning large portions of Asia (Map 2), has provided valuable information on biotype variation in *Nilaparvata lugens*. The intricate network of testing sites in this nursery has unveiled major genetic differences between brown planthopper populations in South Asia—Bangladesh, India, and Sri Lanka—and the rest of Asia (Seshu and Kauffman, 1980). Less dramatic genetic variation occurs in brown planthopper populations within South Asia and other parts of the destructive insect's range in Southeast and East Asia. The International Rice Brown Planthopper Nursery and the International Rice Bacterial Blight Nursery have revealed that brown planthopper biotypes and bacterial blight strains

Figure 9. One of the sites of the International Sorghum Stem Borer Nursery at Mbita Point, Kenya, May 1983. This specialized nursery is coordinated by the International Crops Research Institute for the Semi-Arid Tropics, which is headquartered near Hyderabad, India. The Nairobi-based International Center of Insect Physiology and Ecology collaborates with ICRISAT in this nursery.

are more virulent in South Asia than in East and Southeast Asia (IRRI, 1986). Rice breeders now realize that resistance to brown planthopper or bacterial blight bred into a rice variety in one region may not hold up elsewhere.

The International Pearl Millet Disease Resistance Testing Program (IPMDRTP), coordinated by ICRISAT in India, also provides valuable information about genetic differences in pest populations. This program, which started in 1976, is composed of four subnurseries[1] that screen pearl millet (*Pennisetum typhoides*) for resistance to downy mildew, rust, smut, and ergot and monitor population variations of these pathogens. Several entries in the International Pearl Millet Downy Mildew Nursery (IPMDMN), for example, react differently to *Sclerospora graminicola* at different locations in West Africa, suggest-

1. IPMDMN involves up to twenty sites in India, Senegal, Niger, and Nigeria. IPMRN spans seven locations in India. IPMSN covers seven locations in India, Senegal, and Niger, and the International Pearl Millet Ergot Nursery uses five sites in India and Senegal.

Map 2. Screening sites for the International Rice Brown Planthopper Nursery (information from Seshu and Kauffman, 1980)

ing qualitative differences in populations of the fungal pathogen (ICRISAT, n.d.).

A major advantage of international nurseries is that breeders can get a reading on the performance of materials in a wide range of environments. International nurseries can cover a far greater range of soil and climate conditions than can normally be accomplished at the provincial or national level. This characteristic is especially advantageous to specialized nurseries when the number of suitable sites in any one country may be severely limited. International nurseries allow breeders to spot readily the lines that show wide adaptability and high levels of resistance and to recommend good yielders with greater confidence.

Considerable attention to detail is required to ensure the success of international nurseries. Entries from international centers and national programs must be brought together, inspected for defective or diseased material (Figure 10), cleaned, and packaged. Each trial contains a specific set of instructions as to how the material is to be planted, recommended treatments, and number of replications. Cardboard boxes containing the trials are then sent by air freight to participating countries.

After nursery trials have been planted and harvested, the coordinator monitors returns for accuracy and sends query letters to those who have not reported. Interim reports are sometimes sent to let participants know the outcome of early returns so that they can decide what mix of nurseries they want for the following year and immediately incorporate outstanding material into their breeding programs. Because of the complex logistics of preparing and collating results of nurseries, international agricultural research centers typically coordinate regional or global nurseries.

INGER, for example, involves more than eight hundred scientists and technicians working at some six hundred sites in approximately eighty countries. The number of countries involved in INGER grew dramatically shortly after it started in 1975, but now some fifty to sixty countries normally participate in any given year. Between twenty-five and twenty-eight yield, observational, and specialized nurseries are included in INGER each year (Table 9). In association with the World Meteorological Association (WMO), IRRI recently established a rice-weather nursery to check the performance of certain rice varieties at stations in Asia with high-quality weather records (Greenland et al., 1987). The effect of weather variables on rice yields in a variety of environments can now be gauged reliably.

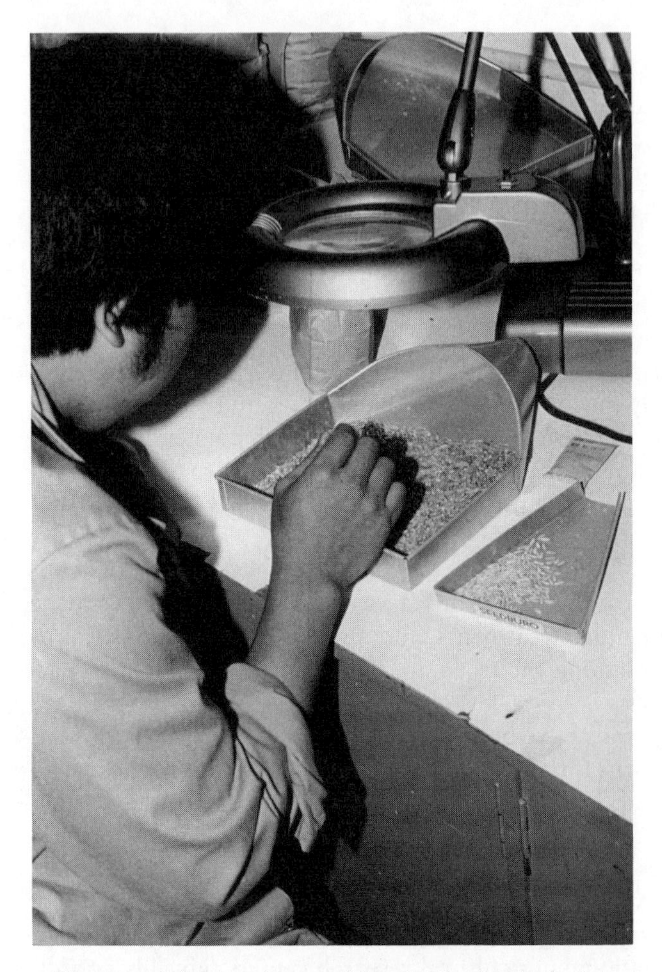

Figure 10. Selecting clean seed for distribution in the International Network on Genetic Enhancement of Rice at the International Rice Research Institute, Los Baños, Philippines, June 1986.

INGER has been the vanguard for spreading modern rice varieties in the tropics. By 1979, just four years after its inception, INGER had launched 30 high-yielding rice varieties in eighteen African, Asian, and Latin American countries (Kauffman et al., 1982). By the end of 1987, INGER had served as the springboard for 161 varieties in forty-seven countries (D. V. Seshu, pers. comm.). Considerable credit for this impressive output rests with national programs that screen the performance of entries and contribute many of their own materials.

INGER is truly a collaborative effort and is not just a springboard

Table 9. Nurseries within the International Network on Genetic Enhancement of Rice

	Irrigated
Yield	
IRYN–VE	International Rice Yield Nursery–Very Early
IRYN–E	International Rice Yield Nursery–Early
IRYN–M	International Rice Yield Nursery–Medium
Observational	
IRON–VE	International Rice Observational Nursery–Very Early
IRON–E	International Rice Observational Nursery–Early
IRON–M	International Rice Observational Nursery–Medium
	Rainfed upland
Yield	
IURYN–E	International Upland Rice Yield Nursery–Early
IURYN–M	International Upland Rice Yield Nursery–Medium
Observational	
IURON–E	International Upland Rice Observational Nursery–Early
IURON–M	International Upland Rice Observational Nursery–Medium
	Rainfed lowland
Yield	
IRRSWYN–E	International Rainfed Rice Shallow Water Yield Nursery–Early
IRRSWYN–M	International Rainfed Rice Shallow Water Yield Nursery–Medium
Observational	
IRRSWON–E	International Rainfed Rice Shallow Water Observational Nursery–Early
IRRSWON–M	International Rainfed Rice Shallow Water Observational Nursery–Medium
IRDWON	International Rice Deep Water Observational Nursery
IFRON	International Floating Rice Observational Nursery
ITPRON	International Tide-Prone Rice Observational Nursery
	Specific stresses
Temperature	
IRCTN	International Rice Cold Tolerance Nursery
Soil	
IRSATON	International Rice Salinity and Alkalinity Tolerance Observational Nursery
Acid upland	Acid Upland Soils Screening Set
Acid lowland	Acid Lowland Soils Screening Set
Diseases	
IRBN	International Rice Blast Nursery
IRBBN	International Rice Bacterial Blight Nursery
IRTN	International Rice Tungro Nursery
Insects	
IRBPHN	International Rice Brown Planthopper Nursery
IRWBPHN	International Rice Whitebacked Planthopper Nursery
IRSBN	International Rice Stemborer Nursery
Nematodes	
IRUSS	International Rice Ufra Screening Set

for IRRI materials. Initially, INGER contained mostly IRRI materials, but the network has matured and exemplifies horizontal transfer of technology between developing countries (Appendix 4). In 1985, national programs contributed 70 percent of the entries, up from 30 percent in 1975, when the network began (Chang et al., 1982; Mashler, 1985). The sizable contribution by national programs indicates their strong interest in the network. National programs are clearly motivated by INGER's tangible benefits. As of 1984, entries from thirty-seven countries had resulted in the release of 77 varieties (IRRI, 1985a). INGER has provided an effective mechanism for the exchange of materials whereby a variety developed in one country can be tested and released in another (Singh, 1985).

Sometimes INGER entries are released directly to farmers in participating countries. Burma has released 20 varieties originating from IRRI and six national programs in the INGER network. Vietnam has launched 12 varieties developed by IRRI and three national programs that were spotted in INGER trials. Gama 318, a rice variety developed by Indonesia, was released as Avinash in Karnataka State, India, in 1985 (IRRI, 1985b). INGER introduced Jaya, an Indian rice variety, to Côte d'Ivoire, Mali, and Senegal, where it has been widely adopted (Pande and Seetharam, 1980). A rice-breeding line from Thailand, SPR 6726-134-2-26, was observed in an INGER plot in Mexico and released in that country in 1981 as Cardenas A80 (S. Sarkarung, pers. comm.). Although Cardenas A80 is no longer in commercial production because of its susceptibility to blast disease, the Thai line made a contribution, albeit briefly, to Mexican agriculture.

INGER entries are frequently used for further crossing, particularly in countries with a strong agricultural research capacity. Thailand's rice program, located at several research stations in many different environments, has benefited greatly from participation in INGER. Of the 24 rice varieties released by the Thai rice program between 1969 and 1981, 11 contain IRRI lines in their pedigree (Dalrymple, 1986). For example, RD21 (Figure 11), which was developed at the Suphanburi Rice Experiment Station and released in 1981, contains IRRI's IR26 in its genetic makeup. Breeders working on RD21 used IR26 as a source for semidwarfism and resistance to biotype 1 of brown planthopper (RRI, n.d.). RD21 has proved popular with farmers because it matures early (125 days) and is well suited for irrigation during the dry season. RD23, another high-yielding variety developed by Thailand's rice program and also released in 1981, has IR32 in its background. The latter IRRI line was a source of semidwarfism and earliness (120

Figure 11. RD21, a semidwarf, high-yielding rice variety developed at the Suphanburi Rice Experiment Station, Thailand. IR26, produced by the International Rice Research Institute in the Philippines, was a source for short stature and resistance to biotype 1 of brown planthopper in the development of RD21.

days). RD21 and RD23 account for most of the irrigated rice area during Thailand's dry season.

INGER's successes are tangible and can be attributed in part to inspired and efficient leadership and governance at the executive level. IRRI provides the services of D. V. Seshu, a senior scientist and plant breeder, as the full-time coordinator to oversee the diverse network. It is doubtful that a part-time coordinator could accomplish such an ambitious task. Resources, both at coordinating nodes and at the numerous testing locations, are in place and reliable. Participants are motivated to make land and labor available for screening germplasm. Finally, the program is well conceived and well managed. Regional committees oversee the delivery of nurseries, thus speeding the arrival of INGER trials into the hands of breeders around the globe (Seshu, 1988).

CIMMYT also operates a far-flung series of international nurseries to test germplasm of its mandated crops—wheat, triticale, barley, and maize (Table 10). As with INGER, the number of countries participating in CIMMYT's nurseries fluctuates each year. In 1974, for example, CIMMYT sent trials of wheat, triticale, and barley to 83 countries; in 1979 and 1980, 134 and 101 countries, respectively, participated (Swanson, 1975:11; CIMMYT, 1981).

The International Wheat Nursery System administered by CIMMYT contains as many as fifty nurseries (Table 10). Several widely planted nurseries, including the International Bread Wheat Screening Nursery (IBWSN), the Elite Selection Yield Trial (ESYT), the Regional Wheat Yield Trial (RWYT) in the Middle East, and the International Spring Wheat Yield Nursery, screen wheat lines for superior yield. IBWSN has been operating since the late 1970s and is currently planted at over two hundred sites on every continent. International nurseries involve a blizzard of acronyms, particularly to the uninitiated, but to breeders they are familiar, and welcome, sources of new material for their programs.

Table 10. Participants in the International Wheat Nursery System coordinated by CIMMYT in El Batán, Mexico, 1977–1987

Year	Nurseries	Cooperators	Number of countries	Seed sent (metric tons)
1977	33	274	97	4.5
1978	31	281	107	9.3
1979	38	307	115	9.3
1980	37	257	101	7.7
1981	35	272	103	8.6
1982	50	280	98	8.6
1983	41	239	91	6.7
1984	45	288	93	7.8
1985	36	312	95	6.9
1986	42	333	98	7.6
1987	47	272	87	8.0

Note: A few nurseries are for screening barley or triticale germplasm, but most are devoted to bread and durum wheats.

Many of the wheat nurseries organized by CIMMYT involve U.S. and European institutions as active partners. For example, the International Winter × Spring Wheat Screening Nursery (IWSWSN), which is designed to exchange desirable qualities between the two wheat groups, depends on assistance from a consortium of U.S. universities. CIMMYT concentrates on transferring sought-after traits from winter wheat to spring wheat, while Oregon State University focuses on introducing certain spring wheat traits to winter wheat. A major objective is to transfer leaf and stem rust resistance from spring wheat to winter wheat (Dubin and Rajaram, 1982). The University of Nebraska tests varieties entered into the nursery for their nutritional attributes. In 1978, ninety-seven breeding programs in forty-eight countries participated in IWSWSN.

CIMMYT also employs several specialized nurseries to check the performance of wheat lines in the face of disease pressure, particularly from rust pathogens. The Regional Disease and Insect Screening Nursery (RDISN), for example, involves more than thirty countries and is administered from Cairo, Egypt, in cooperation with the International Center for Agricultural Research in the Dry Areas. RDISN is used for the early detection and identification of resistance to diseases by screening at "hot spot" locations where diverse races of pathogens are present and evolving. The Latin American Disease and Insect Screening Nursery (VEOLA) is administered from Quito, Ecuador. Lines sown in RDISN and VEOLA are obtained from national programs and are the most disease-resistant and agronomically superior available. CIMMYT makes lines available to national programs that exhibit the most broadly based resistance in these specialized disease nurseries (Dubin and Rajaram, 1982).

Two other specialized disease nurseries operated by CIMMYT help stabilize wheat yields. The Regional Disease Trap Nursery (RDTN) and the Latin American Rust Nursery (ELAR) detect changes in pathogen virulence, thereby providing advance warning to extension services and breeders of the need to change varieties. Although CIMMYT plays a leading role in organizing and administering many wheat disease nurseries in developing countries, other institutions are also involved in operating such specialized nurseries for wheat. The USDA, for example, has operated the International Spring Wheat Rust Nursery, an excellent source of disease-resistant germplasm for wheat-breeding programs in the Third World, for over three decades. These specialized disease nurseries provide national programs with three to five years' lead time between the detection of a new race of a pathogen and when it becomes pandemic.

More than four hundred modern wheat varieties with CIMMYT germplasm have been released by national programs, mostly in the Third World. Many of these materials were obtained through the CIMMYT International Wheat Nurseries, which are planted at over 150 sites. This extensive nursery network is composed mainly of CIMMYT materials, although some entries from national programs are also included (CIMMYT, 1986). Semidwarf wheats are grown on more than 50 million hectares in the developing world (Dalrymple, 1985; Breth, 1986). About 80 percent of the wheat lands in Latin America and Asia, except for China, are now covered by modern varieties. In Sub-Saharan Africa the proportion is 50 percent, and nearly all of Mexico's wheat-growing area is dominated by semidwarfs (Dalrymple, 1985; Breth, 1986). The release of semidwarfs and the use of fertilizer and irrigation have led to a doubling of wheat yields in many countries, particularly India, Mexico, and Pakistan.

More than eighty countries participate in the CIMMYT-coordinated International Maize Improvement Network (Sprague and Paliwal, 1984), which consists of three levels of testing: International Progeny Testing Trials (IPTTs), Experimental Variety Trials (EVTs), and Elite Variety Trials (ELVTs).[2] IPTTs are sent to some twenty-five countries,[3] whereas between seventy and ninety countries participate in EVT and ELVT trials. Promising material is saved at each step for the next testing cycle. The ELVTs are the last stage in the network program; it is then up to national programs to release desirable material directly or use it for further crossing.

Disease outbreaks are one of the primary causes of fluctuating maize yields worldwide so CIMMYT has organized several specialized maize nurseries that screen germplasm for resistance to specific pathogens. In 1974, CIMMYT forged collaborative research projects with several national programs to test maize lines for resistance to downy mildew, corn stunt, and streak virus (Vasal et al., 1982; Sprague and Paliwal,

2. The IPTT is composed of 250 full-sib maize families from each advanced population and 6 check varieties. Based on across-site analysis, 80 to 100 families are selected to regenerate the population for the next stage. In addition, the superior 10 families from each IPTT site and the 10 best across-site families are identified from IPTT data and used to develop experimental varieties, which are sent to cooperators for the next improvement cycle. After information from the experimental variety trials has been analyzed at CIMMYT, superior material is selected for the elite variety trial (Sprague and Paliwal, 1984).

3. In 1988, for example, CIMMYT sent IPTT trials to twenty-one countries (Brazil, Chile, China, Congo, Costa Rica, Dominican Republic, Ecuador, Egypt, Ethiopia, India, Indonesia, Kenya, Mexico, Pakistan, Panama, Peru, Philippines, Turkey, Vietnam, Zaire, and Zambia).

1984). A shuttle breeding program has been set up so that growing cycles are alternated between cooperating countries and CIMMYT facilities in Mexico. In Mexico, breeding concentrates on incorporating agronomically desirable features into maize lines showing some indication of disease resistance. Thailand and the Philippines collaborate with CIMMYT to develop varieties resistant to downy mildew, and El Salvador and Nicaragua cooperate in screening maize germplasm for resistance to corn stunt. Tanzania and several West African nations work with CIMMYT on streak resistance.

As of 1981, 108 maize varieties drawn from the International Maize Improvement Network had been released in twenty-six countries (Sprague and Paliwal, 1984). By 1986, forty-three national programs had released 147 open-pollinated maize varieties or hybrids derived from materials entered in the network. These generally shorter, modern varieties usually produce 20 to 35 percent better yields than slower-maturing traditional varieties (Borlaug, 1983).

Brazil has released 12 varieties originating in the International Maize Improvement Network, mostly on the basis of high levels of disease resistance. Guatemala, where maize is a basic staple, has launched 10 varieties and 3 hybrids stemming from this nursery; these varieties and hybrids occupy about 40 percent of the country's maize-growing area. In 1985–86 approximately half of 80 million hectares in developing countries was planted with commercial seeds of improved hybrids and improved varieties (Timothy et al., 1988). About half of the crop area planted to improved materials is in China, Brazil, and Argentina so about 20 percent of the total maize areas of developing countries is planted with commercial varieties or hybrids. Many of these new materials have CIMMYT germplasm in their pedigrees.

Nurseries are particularly well developed for the major cereals, but networks for germplasm testing have been set up for other food crops such as pigeonpea (*Cajanus cajan*), soybean, common bean (*Phaseolus vulgaris*), and groundnut. ICRISAT, for example, coordinates specific stress nurseries to screen groundnut for resistance to rosette virus, mottle virus, and bud necrosis. The nursery for rosette virus involves sites in Georgia (United States), Scotland, Germany, Niger, Nigeria, and Malawi. The nursery for groundnut mottle virus incorporates sites in Georgia, China, Thailand, Philippines, and Australia, and the bud necrosis nursery spans Texas, Brazil, France, India, China, Thailand, and Australia. In addition to field screening, these specialized disease nurseries for groundnut involve a range of research activities, including genetic engineering.

The International Bean Yield Adaptation Nursery (IBYAN; Table 8), a global enterprise coordinated by CIAT, has exposed bean breeders in more than thirty countries to a rich assortment of common beans since 1976. As of 1983, IBYAN had launched 48 varieties in developing countries. All materials developed through IBYAN resist bean common mosaic virus, and IBYAN has helped locate lines resistant to all known races of the pathogen *Colletrotichum lindemuthianum*, which cause anthracnose of beans, a worldwide and highly destructive bean disease. In collaboration with CIAT, Guatemala's Instituto de Ciencia y Tecnología Agrícola (ICTA) has developed three new bean varieties that have helped the country become self-sufficient in bean production (Correa, 1986). CIAT, based near Cali, Colombia, received the CGIAR King Baudouin Prize for Agricultural Development for its role in turning Guatemala's bean production around. Cuba's bean output has also increased significantly as a result of its involvement in IBYAN.

More than 130 Third World nations joined networks to test crop-breeding material during the 1970s (Hanson, 1979). International nurseries are clearly able to surmount ecological, religious, ethnic, and language differences between nations and have accelerated the launching of high-yielding varieties, particularly in the Third World. In 1980, the additional wheat and rice produced by green revolution technologies were worth an estimated $56 billion, of which $10 billion was attributable to the improved genetic potential of the new varieties (Wolf, 1986).

Agricultural Machinery Networks

Several networks designed to promote the use of relatively simple farm implements and intermediate technology machinery have been established, principally in Asia. These networks draw together engineers and manufacturers working with low-cost threshers, reapers, pumps, plows, power tillers, weeders, rice transplanters, seed and fertilizer applicators, and grain cleaners. Such machines are designed to be made easily by local foundries and blacksmiths. Machines being tested in the various networks can be manually operated, although some require a small motor or, in rare cases, a tractor. Small motors under ten horsepower are readily available in developing countries, are easy to maintain, and are cost-effective in many farming situations. Transplanters and harvesters that are tested in agricultural research networks can often be pulled by livestock in the event that tractors are not economically feasible.

Agricultural machinery networks are relatively uncommon, in part because the private sector develops so much agricultural machinery. Little international collaboration has evolved in the planning and development of farm machines and implements, even among countries in which the public sector is primarily responsible for designing and producing agricultural machinery. Two of the oldest and probably best-developed agricultural machinery networks are discussed here, and they are both independently planned networks.

The IRRI Industrial Extension Network has focused on machinery for rice farmers who till from two to ten hectares of land. The main purpose of the network, which started in 1976 and essentially phased out in 1985,[4] was to disseminate machinery designs developed by IRRI engineers to small-scale manufacturers in Thailand, Indonesia, India, Burma, the Philippines, and Egypt (Bockhop et al., 1985). Network participants also tested some local designs. The extension services in participating countries were asked for advice at the blueprint stage and for assistance in disseminating prototypes.

More than five hundred manufacturers have been involved in one or more IIEN projects (Figure 12). Manufacturers were given blueprints at no cost and were encouraged to produce items with or without modification. To generate interest and reduce risks to manufacturers, a government agency often placed an initial order for a machine or implement for demonstration purposes. Workshops held every two to three years at IRRI headquarters in the Philippines and training courses with field trips were also integral to IIEN. Instruction was given to farmers as well as engineering trainees. In Thailand, for example, more than fifteen hundred farmers attended four sessions to learn how to operate and maintain inclined plate planters (Bockhop et al., 1985). A network newsletter alerted manufacturers and extension agents to implements and machines that were gaining acceptance in particular environments. Such information helped manufacturers and engineers struggling with similar problems and may have prevented some from "reinventing the wheel."

Paddy rice cultivators in tropical Asia have been the main customers for machines and implements promoted by the network. Some six hundred thousand machines and implements stemming from IIEN activities have been made in ten countries. In India, sixty-four cooper-

4. Reasons for the winding down of IIEN are discussed in Chapter 11.

Figure 12. A small independent machine shop that participated in the IRRI Industrial Extension Network, Los Baños, Philippines, June 1986.

ating manufacturers fabricated over ten thousand hand tractors, axial-flow pumps, transplanters, threshers, and dryers between 1979 and 1984 (Bockhop et al., 1985). Although this figure is small compared to the number of farmers in India, many of the machines serve several farms.

IRRI-designed axial-flow threshers have been the most successful machines promoted by the network. Several designs of axial-flow threshers have been built by IRRI and collaborators in IIEN, all of which are powered by small, easily maintained motors and can be hauled to fields by livestock, tractors, or a work crew. IRRI-designed axial-flow threshers are presently manufactured in eight countries; at least four hundred rural enterprises are turning out the threshers in the Philippines and Thailand alone. King Carl Gustaf of Sweden presented the International Inventors Award to Amir Khan of IRRI in 1986 for developing the axial-flow thresher. The award was established to recognize outstanding innovations in the fields of water, energy, forestry, and industry.

Another agricultural machinery network spanning Asia, the Regional Network for Agricultural Machinery, is not strongly linked to an international center as is IIEN. And, unlike IIEN, which concentrates on technology for small-scale rice farmers, RNAM's activities involve a large range of implements for cash and food crops. Further, RNAM caters to the needs of small-, medium-, and large-scale farmers.

With a small secretariat provided by the U.N. Economic and Social Commission for Asia and the Pacific (ESCAP) in Los Baños, Philippines, RNAM links ten research institutions in developing countries with more than two hundred small-scale manufacturers in Bangladesh, India, Indonesia, Iran, Nepal, Pakistan, Philippines, South Korea, Sri Lanka, and Thailand.

RNAM's policy and operating procedures are determined by a governing body composed of senior-level representatives from participating countries and donor organizations which meets once a year. The governing body decides on matters relating to financial contributions, external assistance, and intercountry cooperation. A technical advisory committee also meets annually to formulate the work program, evaluate progress, and provide input on engineering and other technical issues.

In addition to organizing workshops and annual ten to twelve-week training courses on different themes, RNAM pays for the shipping costs of machine prototypes of interest to participating institutions (Table 11). Institutions or companies interested in acquiring machines for testing pay for the actual cost of the product and any processing or handling fees at the port of entry. This system enables institutions to

Table 11. Status of prototypes of agricultural machinery and implements exchanged between RNAM participants as of April 1986

Status	Number of prototypes
Initial testing	9
Extensive testing	7
Modification	4
Unsuitable	15
Popularization	1
Commercial production	1
In transit	5
Lost in transit	4
Unable to supply	7
Total received	37

Note: Some categories overlap.

try out potentially useful machines and implements more quickly and cheaply than if they had to design and construct them.

Several machines or implements shared among RNAM participants have been put into commercial production or are on the verge of adoption. A reaper-windrower developed in Pakistan is now sold in India, and a semiautomatic sugarcane planter created in India is being actively promoted in Pakistan. RNAM sometimes helps obtain promising prototypes from countries outside the network. Thanks to the offices of the network's secretariat, for example, a Chinese vertical reaper is now being manufactured in commercial quantities in Pakistan (RNAM, 1985).

Outlook

Material exchange networks play an essential role in international evaluation of research products, particularly crop germplasm and agricultural machines. Many international nursery networks are moving toward collaborative research, and this trend is likely to continue as the early emphasis on testing begins to evolve into more detailed studies of specific diseases or insect problems or problems relating to crop adaptation to the growing environment.

8

Scientific Consultation Networks

Our survey of scientific consultation networks covers those cooperative research efforts in which many individual programs were already under way when the network was formed. Scientific consultation networks are a mechanism for sharing ideas on a major problem or issue, learning about new methodologies, and making minor changes to ongoing research projects. The participants mostly facilitate each others' programs rather than work together on a joint product, as in collaborative research networks. Most research efforts in scientific consultation networks are independently planned by participants, and only occasionally do they involve monitoring tours. Unlike collaborative research teams, scientific consultation networks are not always based on a founding document that explores a problem and establishes a need for closer unity among research programs in solving a particular problem.

Scientific consultation networks can be set up relatively quickly because no major realigning of research programs is necessary, at least during the early phase. Consequently, these networks are fairly common, spanning a vast array of topics from livestock nutrition to biological nitrogen fixation (Table 12). It should be kept in mind, however, that not all networks fit conveniently into categories. Some scientific consultation networks are transitional to true collaborative research ventures, while others are similar to material or information exchange networks.

No network typology can be purely objective, and our four-part typology calls for subjective judgment. We have perhaps erred on the

Table 12. Some scientific consultation networks in international agricultural research

Network	Research area	Region	Training	Tours*
Amelioration de la Culture du Mais	Maize improvement	Africa, Latin America	Yes	Yes
Amelioration de la Culture du Riz	Rice improvement	Africa	Yes	No
ARNAB	Livestock nutrition	Africa	Yes	No
CIMMYT/ESA	Farming systems	East Africa	Yes	No
East African Cowpea Improvement Network	Cowpea breeding and agronomy	East Africa	Yes	Yes
Eastern Africa Regional Bean Research Project	*Phaseolus* bean improvement and production	East Africa	Yes	Yes
INTSOY	Soybean processing for human and livestock consumption	World	Yes	No
MIRCENs	Microorganisms in BNF†, industrial fermentation, pest control	World	Yes	?
ONESASA	Improvement of annual oil crops	Africa, South Asia	Yes	Yes
PESTNET	Management of crop pests and disease vectors of livestock	Africa	Yes	Yes
RISPAL	Animal production	Latin America	Yes	Yes
RRS	Agronomic measures to counteract moisture stress in crops	Africa, Canada, France	Yes	No
SADCC Cowpea Improvement Network	Cowpea breeding and agronomy	Africa	Yes	Yes
SADCC/CIAT Southern Africa Regional Bean Research Project Network	Bean improvement	Africa	Yes	Yes
Small Ruminant and Camel Group Research Network	Improvement of productivity of camels and small ruminants	Africa	Yes	No
SUAN	Farming systems and management of natural resources	Southeast Asia	No	No
WAFSRN	Farming systems	West Africa	No	No
West and Central African Cowpea Improvement Network	Cowpea breeding and agronomy	West and Central Africa	Yes	Yes
West Africa Groundnut Collaborative Network	Groundnut improvement	West Africa	Yes	No

Note: For the larger networks, see Appendix 1 for network acronyms, coordinators, starting dates, and number of participating nations.

*Regular monitoring tours, composed of network participants, to research sites.

†Biological nitrogen fixation.

side of simplicity in using only four major network categories, but a highly complex typology can be unwieldy and confusing. We do not wish to imply that the scientific consultation networks discussed here could not also fit into one of the other categories. The criteria we used to place them included their major business, whether research efforts were independent or tightly linked, and whether research was jointly planned.

Rather than review all scientific consultation networks, we have selected a few examples from a wide range of agricultural research fields. This chapter portrays the rich tapestry of networking in agricultural research, identifies common threads in networking, analyzes the uneven performance of cooperative research efforts, and sifts out specific dividends from such ventures.

Farming Systems

Agricultural research can benefit directly or indirectly from studies of farming systems that seek to identify constraints to increased agricultural production, whether of a biological, agronomic, cultural, or socioeconomic nature. Farming systems research seeks to understand farmers' difficulties and to improve their circumstances and can result in sounder recommendations to farmers and more appropriate research priorities than was previously available (Collinson, 1987).

A multidisciplinary team generally undertakes diagnostic surveys to uncover principal stumbling blocks to improved crop and livestock yields (Lamung, 1985). After the major problems have been identified, the team may undertake more in-depth research or the task may be taken up by relevant specialists. A central idea in farming systems research is to provide feedback to crop breeders, agricultural engineers, fertilizer specialists, and agricultural extension agents on criteria farmers will use in evaluating new technologies.

Several scientific consultation networks focusing on farming systems work arose in the 1970s and early 1980s to address the growing need for more adaptive research. Africa, where the need to increase agricultural production is particularly urgent, is home to two of them: the CIMMYT Eastern and Southern Africa Economics Program (CIMMYT/ESA),[1] based in Nairobi, and the West African Farming Systems Research Network.

1. Originally known as the CIMMYT Eastern Africa Regional Economics Program. See Appendix 2 for more information on CIMMYT.

The main purpose of CIMMYT/ESA, which encompasses fifteen countries from Sudan in the north to Swaziland and Lesotho in the south, is to promote systems-based, on-farm research techniques in national agricultural research systems. The network conducts training courses, acts as a clearinghouse for methodological issues in farming systems research, and provides opportunities to compare problems and accomplishments among different countries. When the network began in 1976, CIMMYT economists concentrated on demonstrating on-farm research techniques, especially diagnostic surveys and planning experiments (M. Collinson, pers. comm.). As interest in farming systems research grew, particularly from 1979 onward, training was included in the regional program, and by 1983 it dominated the network's activities. This shift in emphasis was warranted by the success of initial efforts to raise awareness of the need for direct interaction between small farmers and scientists. It demonstrates the importance of operating with a flexible strategic plan.

CIMMYT/ESA has done more to promote farming systems research in Africa than any other single entity (Fresco and Poats, 1986). Its catalytic effect has been most pronounced in Zambia, Malawi, Zimbabwe, and Ethiopia, where national agricultural research and extension programs have been reorganized to accommodate farming systems research. In Zambia, Malawi, and Ethiopia diagnosis and on-farm experimentation are now routine, and links between farmers, researchers, and extension workers are reinforced by team visits to farms by technical and social scientists.

A more recent farming systems network established in West Africa is in the early stages of transition from an information exchange operation to an independently planned research consultation network. The West African Farming Systems Research Network, also known as RESPAO (Réseau d'Études des Systèmes de Production en Afrique de l'Ouest), started informally at a meeting held in 1982 at IITA in Ibadan, Nigeria. Participants at that meeting decided to form a network to link farming systems researchers in ten West African nations, to compile an inventory of terminology in farming systems research, to conduct a literature search on the subject relevant to West Africa, and to map agroecological/farming systems zones in the region. A newsletter was also launched to inform participants of the network's progress.

Much work remains to be done before WAFSRN completes the transition to a collaborative research network. Still, some ground has been broken and interest in the network is clearly growing since

WAFSRN membership had nearly doubled to seventeen countries[2] by 1986. WAFSRN's first annual workshop held in Dakar, Senegal, in 1986 attracted fifty delegates from all seventeen member countries.

The Semi-Arid Food Grain Research and Development Project (SAFGRAD), based in Ouagadougou, Burkina Faso, was chosen as the new coordinating body for WAFSRN at the 1986 workshop. A full-time coordinator, an experienced Senegalese scientist, came on board in 1987 to coordinate WAFSRN from a base in SAFGRAD. A steering committee, containing a maximum of seven scientists, has been established and a research plan drawn up. The research agenda is envisaged in three phases. In the first phase, the network will strive to provide an inventory of institutions and researchers involved in farming systems research in the region, assess training facilities, and identify sources of technical and financial assistance.

In the medium term, WAFSRN hopes to devise a valid zoning of West African farming systems, to facilitate technical backstopping of farming systems research in the region, and to advise in the recruitment of farming systems researchers. In the third phase, WAFSRN is expected to emerge as a full-fledged collaborative research network. WAFSRN's long-term goal is to help design and implement a coordinated regional farming systems research program.

Livestock Nutrition

Livestock networks are not nearly as common as crop or natural resource–oriented cooperative research ventures, but several organizations are uniting efforts across countries to upgrade meat, milk, and fiber yields, as well as livestock traction power for farm chores. Africa is the setting for several international livestock research networks. Networking in African livestock research is spurred by the need to increase the protein intake of the continent's rapidly growing human population and to improve endurance and resilience of cattle performing agricultural tasks. Livestock are a major source of income for many Africans. Domesticated animals contribute between 20 and 30 percent of the gross domestic product in most Sub-Saharan countries.

2. The countries participating in WAFSRN as of 1986 were Benin, Burkina Faso, Cameroon, Cape Verde, Côte d'Ivoire, Gambia, Ghana, Guinea, Guinea-Bissau, Liberia, Mali, Mauritania, Niger, Nigeria, Senegal, Sierra Leone, and Togo.

Furthermore, pastoralism is still a way of life for close to 40 million people (Hesterman, 1986; Simpson and McDowell, 1986).

One of the several livestock research networks coordinated by the Addis Ababa–based International Livestock Center for Africa focuses on better ways to incorporate crop by-products into feed for cattle, sheep, goats, and pigs. The need to stretch or substitute imported livestock feed is particularly important in developing countries because the consumption of feed grain is rising dramatically. In the Third World, the use of cereals for livestock feed is increasing 50 percent faster than direct consumption of grain by humans, and concern is mounting that grain prices will rise, adversely affecting the poor (Sarma, 1986). To help stem the tide of grain importation for livestock by using domestic plant products more effectively, the African Research Network for Agricultural Byproducts was launched in 1981 with support from the Australian government.

ILCA was chosen as the hub for ARNAB because of its expertise in livestock management systems and superior library and printing facilities. The network is governed by an eight-member steering committee composed of four ex-officio members (the head of the nutrition unit at ILCA, the ARNAB coordinator, the coordinator of the Pastures Network for Eastern and Southern Africa, and the program officer for crop and animal production systems of IDRC, Nairobi) and four regional representatives from different parts of Africa, who are elected for two-year terms at annual workshops.

ARNAB serves as a mechanism to share information and to foster research on crop by-products as livestock feed. Its strategic plan has outlined eight main tasks in the short to medium term. First, the network is designed to stimulate and strengthen research on crop residues and agroindustrial use of crop by-products through cooperative research. Second, ARNAB is to collect, analyze, and disseminate literature on the use and processing methods of crop by-products. Third, critical reviews of the literature on crop by-products are to be prepared and disseminated among network members. Fourth, standard evaluation methods and terminology are to be developed. Fifth, participants are to collect samples of agricultural by-products and to conduct quantitative surveys. Sixth, technologies are to be developed to improve the nutritional value of crop by-products. Seventh, training is to be arranged at the technical and master's degree levels, and finally a workshop proceedings and a newsletter are to be published. English and French versions of the quarterly ARNAB newsletter are currently sent to six hundred individuals and institutions.

Specific research projects currently under way within the network include mixing groundnut residues with livestock feed in Senegal, incorporating cacao pods with livestock meal in Nigeria, and employing maize stover for livestock feed in Cameroon. An ILCA scientist visits network colleagues in Senegal, Nigeria, and Cameroon twice a year to discuss research progress and problems. Cyprus was to join ARNAB, but funding difficulties have checked any expansion of the network (IDRC, 1986).

In addition to periodic workshops held at various locations, ILCA offers training for three to six weeks to midlevel technicians. In 1982, eighteen Africans participated in the ARNAB training course at ILCA's headquarters in Addis Ababa. ILCA also serves as a clearinghouse for funding requests from national programs to conduct research on crop by-products.

Outlook

Scientific consultation networks are often started with the aim of becoming collaborative research networks. Some networks may achieve this goal, but for others a major focus on scientific consultation may meet the needs of the participants. Shifting established research programs to a true collaborative mode can be difficult, and not all research problems lend themselves to the guided collaborative model.

9

Collaborative Research Networks

The major distinguishing feature of collaborative research networks is that they are jointly planned and executed. Unlike scientific consultation networks, which are a relatively loose association of preexisting research programs, collaborative research networks usually entail a ground-up design of a new team effort. Where collaborative research networks are the result of the amalgamation of prior individual programs, the latter normally reorganize or change to accommodate common goals. Collaborative research networks usually require tighter coordination, closer management, and stricter adherence to common research methodologies than do scientific consultation networks.

Because research is more integrated in collaborative programs, the roles of participants are generally better defined than in other kinds of networks. Also, collaborative research networks are more likely to entail formal protocols or memorandums of understanding. They tend to have more "add-on" components than do straightforward material exchange networks or scientific consultation efforts. Collaborative research networks normally have regular meetings of participants, a steering committee, a technical advisory committee, as well as monitoring tours and training courses tailored to the network's needs. And because most collaborative research networks are new, ground-up enterprises, a substantial and authoritative founding document is particularly important to help establish priorities and set the research agenda.

Many networks aspire to be collaborative research networks, but they can more realistically be classified as scientific consultation or material exchange networks. Again, we do not wish to imply that all networks should pursue a development path toward a collaborative research program. Each type of network fulfills a need, and every network must find its niche or fade away. One of the great attractions of collaborative research networks is their sense of cohesion, unity of purpose, greater propensity for insight, and high productivity, attributes not exclusive to collaborative research but found less frequently in other types of networks.

Collaborative research networks are operating on every continent in the Third World and embrace national programs in various stages of development. Collaborative research projects tackle agricultural problems ranging from livestock diseases to potato diseases (Table 13). We sample collaborative research networks in various fields, including farming systems (Asian Rice Farming Systems Network), livestock diseases (Trypantolerant Livestock Network), commodity-oriented research (PRECODEPA, SAPPRAD, some CRSPs), and soil fertility and agrotechnology transfer (INSURF, IBSNAT).

Asian Rice Farming Systems Network

The oldest international network dealing with farming systems research is the Asian Rice Farming Systems Network, which started in 1975. ARFSN links research following jointly planned methodologies but not necessarily a common work plan. It was first known as the Asian Cropping Systems Network and then the Asian Farming Systems Network and was renamed the Asian Rice Farming Systems Network in 1983 (Carangal, 1988). The network is guided by the Asian Rice Farming Systems Working Group, which meets annually and is composed of national cropping/farming systems program leaders, the ARFSN coordinator, an economist, and from two to five scientists in Asia who are invited to provide input on specific emerging issues.

ARFSN is coordinated by IRRI and encompasses more than five hundred researchers and technicians in fifteen nations. All the participating countries, except for the Malagasy Republic, are in South or Southeast Asia. Indonesia alone has more than one hundred trained scientists involved with ARFSN. As a result of interactions with ARFSN, the Philippines national program has set up 115 cropping systems research sites around the archipelago nation.

Table 13. Some collaborative research networks

Network	Research area	Region	Training	Tours*
AFRENA	Agroforestry	Africa	Yes	Yes
ARFSN	Rice farming systems	Asia	Yes	Yes
CRSP—Bean/ Cowpea	Breeding, storage, management, processing	Africa, Latin America, United States	Yes	Yes
CRSP—Peanut	Breeding, storage, management, processing, consumption	Africa, Southeast Asia, Caribbean, United States	Yes	Yes
IBSNAT	Transfer of agrotechnology between similar soil types	World	Yes	Yes
INSURF	Rice fertilization trials, including mulches, soil fertility	World	Yes	Yes
IWBP	Epidemiology and control of witches' broom disease of cacao	South America, Trinidad	No	Yes
PRECODEPA	Potato breeding, postharvest technology, diseases, pests, socioeconomics, seed production	Central America, Caribbean	Yes	Yes
SAPPRAD	Potato agronomy, seed production, breeding, postharvest technology, technology transfer	Southeast Asia	Yes	Yes
Trypanotolerant Livestock Network	Epidemiology and control of livestock trypanosomiasis	Africa	Yes	Yes

Note: For network acronyms, coordinators, starting dates, and number of participating nations see Appendix 1.

*Regular monitoring tours, composed of network participants, to research sites.

One reason that the network has caught on so well is that it seeks to improve the productivity of the most important staple crop in a vast, densely populated region. ARFSN scientists are acutely aware that rice growers also cultivate other crops. ARFSN thus emphasizes on-farm evaluation and sharing of new technologies designed to increase the cropping intensity of rice as well as crops grown before and after rice, thereby boosting income and creating more employment opportunities. Other ARFSN goals are to integrate crop production better with livestock and aquaculture and to find more uses for crop residue in manufacturing industries.

Figure 13. A water buffalo participating in feeding trials of the Asian Rice Farming System Network, Malanay village, north-central Luzon plain, Philippines, June 1986.

Crop-livestock integration is a relatively recent component of ARFSN. Asian farmers have used water buffalo and cattle for thousands of years to prepare fields and transport goods, but modern technology development often does not include livestock. Many farmers cannot become fully or even partially mechanized because their landholdings are too small or they have limited access to credit. At two villages near Santa Barbara, Pangasinan Province, in the northern part of the central Luzon plain in the Philippines, ARFSN researchers compare weight gain in water buffalo when fed varying proportions of nitrogen-rich forages and fibrous-crop residues, principally rice straw (Figure 13). Minerals also supplement the diets of the water buffalo. The forages under study are all nitrogen-fixing legumes such as leucaena (*Leucaena leucocephala*), madre de cacao (*Gliricidia sepium*), a common shade tree in coffee and cacao plantations, and sesbania (*Sesbania rostrata*), which thrives in wet environments and has nitrogen-fixing nodules on stems and branches rather than on roots. In addition to providing protein-rich livestock feed, these fast-growing trees provide firewood and green manure.

In Indonesia, ARFSN scientists are conducting controlled experiments with various livestock on five-hectare homesteads at Batumarta, a Sumatra transmigration project near Baturaja. Control families have no livestock, while other study families have varying numbers of chickens, goats, and Bali cattle. The object is to quantify what mix of livestock and crops produces the best net financial and nutritional gain so as to increase household income by at least $1,500 a year (CRIFC, 1986).

Bali cattle are smaller and less powerful than zebu cattle, but they tolerate heat better and thrive on relatively poor forage. Study families in the transmigration area appreciate Bali cattle for their endurance when plowing under the midday sun (Figure 14). The recently introduced brown and white Bali cattle have piqued the interest of numerous families in this pioneer settlement scheme.

ARFSN has forged links with outside research institutions to help develop technology for commodities other than rice (Carangal, 1988). ARFSN has thus reached out to its larger environment to seek expertise and reinforce the collaborative research effort. ARFSN collaborates with CIMMYT on maize and wheat; IITA on cowpea and soybean; ICRISAT on groundnut, sorghum, pigeonpea, and chickpea; CIP on potato and sweet potato; the Asian Vegetable Research and Development Center (AVRDC) on mungbean and soybean; the International Center for Living Aquatic Resources Management (ICLARM) on fish culture; CIAT on forage crops; and the Australian Center for International Agricultural Research on pigeonpea and forage crops.

ARFSN's activities have produced tangible dividends in several participating countries. In Iloilo Province, Philippines, new cropping patterns devised by ARFSN scientists in collaboration with farmers now permit two crops to be grown a year instead of one, thereby increasing farm income by an average of 30 percent (IRRI, 1985c). This intensified cropping pattern has been made possible by direct seeding rather than transplanting and the introduction of quick-maturing, semidwarf rice varieties that can be reaped in about 110 days instead of the 175 days required by traditional varieties.

ARFSN was instrumental in introducing direct seeding of rice to parts of Thailand. In 1975, during the first year of ARFSN, a Thai participant on an ARFSN monitoring tour observed direct seeding of rice by hand and two-row seeders in dry, rainfed soil on Sumatra. He tried this idea in the Pimi area of Northeast Thailand, where it has been widely adopted. This sowing practice is well adapted to the area because seeds can remain dormant until the southwest monsoon arrives in June (D. Chandrapanya, pers. comm.). With traditional transplanting, seedlings are susceptible to drought if the summer monsoon is late.

Figure 14. Bali cattle plowing a field in the Batumarta Transmigration Project, Sumatra, Indonesia, June 1986.

Trypanotolerant Livestock Network

Most livestock research networks deal with nutrition or traction, but one of the best examples of collaborative research on an international scale is provided by the Trypanotolerant Livestock Network in Sub-Saharan Africa. This network is characterized by world-class research and keen participation by highly motivated scientists and technicians in a region torn by political conflict and retarded by struggling research institutions.

Before outlining the scope and research strategy of the Trypanotolerant Livestock Network, which is coordinated by ILCA in close association with the International Laboratory for Research on Animal Diseases, we briefly review the nature of trypanosomiasis and the history of control efforts.

African trypanosomiasis, a debilitating and often fatal livestock disease that occurs in low-lying, tropical portions of the continent, is triggered by various species of blood parasites (*Trypanosoma brucei*, *T. vivax*, and *T. congolense*). These protozoans infect some wild and domesticated animals. Two of the pathogens, *Trypanosoma brucei gambiense* and *T. brucei rhodesiense*, also induce sleeping sickness in people. Tsetse flies transmit the blood parasites and infest over one-fifth of Africa and three-quarters of the subhumid zone (600 to 11,000 mm rain per year), an area harboring sizable livestock populations (Simpson and McDowell, 1986). Some 33 million cattle live in the subhumid zone, slightly over one-fifth of the cattle population south of the Sahara. As a result of the prevalence of trypanosomiasis, livestock in the subhumid zone have the lowest productivity of any livestock populations in the Sub-Sahara. In Africa as a whole, some 50 million cattle, 30 million sheep, and 40 million goats are exposed to trypanosomiasis (ILRAD, 1987).

Traditional control methods for African trypanosomiasis, ranging from eliminating wild animals that serve as reservoirs for the pathogens to clearing brush and trees that shelter the vectors, have had limited success. In some cases, control measures have exacerbated social and environmental problems such as reducing game yields and accelerating soil erosion.

It has long been known that certain African cattle breeds, particularly N'Dama (Figure 15) and West African shorthorn, tolerate trypanosomiasis, but their potential has until recently been largely ignored. These hardy breeds have been considered too small and unproductive for livestock improvement. This prejudice can be traced to the late fifteenth century, when Europeans first set foot in West Africa and remarked that cattle were "runty" and milk production was meager (Crosby, 1986). Furthermore, resistance to the disease was thought to be a local phenomenon and not an intrinsic or genetic trait that would hold up if the cattle were taken to other tsetse-infested areas or that could be passed on to other cattle through breeding.

The roots of the Trypanotolerant Livestock Network extend back to 1977, when UNEP and FAO commissioned a survey of trypanosomiasis in Africa and the distribution of trypanotolerant breeds of cattle, sheep, and goats. A major conclusion of the study, based on intensive fieldwork in eighteen countries, was that trypanotolerant livestock, although generally small, are as productive as other breeds in areas with low rates of trypanosomiasis transmission and much more productive in high-risk environments for the disease (ILCA/FAO/UNEP, 1979). This baseline study provided a believable founding document for the network, which was essential in raising donor support (J. Trail, pers. comm.). It also set the stage for more productive work by network participants.

The survey provided several leads to pursue in trypanosomiasis research, including the potential role of livestock breeds that resist the

Figure 15. N'Dama bull, Yonfolila, Mali. Courtesy of the International Livestock Center for Africa, Addis Ababa.

disease. A collaborative approach was considered the best strategy in following up on promising research avenues because trypanotolerance could be compared more easily over a wide range of environments if participants used the same methodology. An informal, jointly planned research network was thus initiated in 1981 to coordinate research on trypanotolerant livestock in Gabon, Côte d'Ivoire, Nigeria, and Zaire. The Trypanotolerant Livestock Network was formalized in 1983 under the coordination of ILCA's Nairobi office and expanded into Togo, Senegal, and the Gambia. Benin and Congo joined in 1985, and Tanzania, Ethiopia, and Kenya were involved in the network by 1986, bringing the total number of participating countries to ten.

The aim of the network is to improve livestock production through a better understanding of genetic and acquired resistance to the disease as well as ways to improve the effectiveness of current trypanosomiasis control measures such as drug treatments and the use of insecticides and traps to reduce vector populations (ILRAD, 1986b). A strategy of relying on chemotherapy alone is unlikely to be successful in the long run because many livestock owners in Africa cannot afford regular veterinary care and the trypanosome parasites are likely to develop resistance to drugs. Hence the emphasis is on searching for heritable traits that confer some tolerance to the disease.

In addition to coordination, ILCA provides leadership and research expertise in livestock management. In its planning, training, and surveying in the network, ILCA works closely with two other international centers in Nairobi, ILRAD and ICIPE. Much of the network's laboratory work is done at ILRAD's and ICIPE's modern, well-equipped facilities. ILRAD and ICIPE provide expertise in veterinary medicine, parasitology, and entomology, while ILCA covers livestock management, genetics, nutrition, and socioeconomics. The three international centers collaborated to prepare the basic manual on methodology for the network (Murray et al. 1983).

Standardized animal health data collected at the thirteen network sites include verification of trypanosomiasis infection, species of trypanosome, and presence of other diseases such as helminthiasis and anemia (ILRAD, 1986a). Tsetse density is estimated at the sites using traps, and captured flies are dissected to determine trypanosome infection rates and sources of meals. Network participants record data on tsetse challenge, animal health, and livestock productivity in varied environments, from humid forest to savanna.

Information from the thirteen network sites is sent monthly to ILCA's Nairobi office, and if no major omissions or inaccuracies are noted, the data are entered into a microcomputer (ILCA, 1986b). If discrepancies are found, a message is sent by telex or private courier to the field site in question. Once the data are in magnetic form, tests are conducted to detect impossible values. Reliable data are taken on floppy disks to ILCA's headquarters in Addis Ababa for further analysis on a mainframe computer.

Trypanotolerant and susceptible livestock are studied under different management systems, ranging from commercial ranches to traditional and seminomadic households. In 1986, a Trypanotolerant Livestock Network site in Côte d'Ivoire began experimenting with traps designed to reduce tsetse populations. And in Gabon, tests began in 1986 on insecticides effective against tsetse flies that can be incorporated in dips used to control ticks.

N'Dama are studied in villages in the Gambia, Senegal, and Zaire and on two commercial ranches in Zaire (ILRAD, 1986c). In parts of West Africa, information is also gathered on trypanotolerant West African shorthorn cattle, Djallonke sheep, and West African dwarf goats. The performance of susceptible and trypanotolerant breeds is compared at many of the network sites to provide quantified data on the yield advantage of trypanotolerant breeds in areas of trypanosomiasis transmission.

Some 8 million of the humpless N'Dama and West African shorthorn cattle roam West Africa, but they account for only 5 percent of the continent's cattle population. Their contribution is likely to increase in the future following confirmation that trypanotolerant breeds thrive in areas of low to moderate trypanosomiasis transmission and that this resistance is genetic (Murray et al., 1981, 1982; ILCA, 1986b). N'Dama are not immune to African trypanosomiasis, however; in extremely high tsetse challenge environments, even trypanotolerant breeds can suffer.

To further work on intrinsic factors that account for tolerance to trypanosomiasis, ten frozen N'Dama embryos were flown from the Gambia and implanted in Boran cows at ILRAD's headquarters in Nairobi in 1983 (ILRAD, 1986b). The surrogate cows gave birth to ten healthy N'Dama calves in 1984. In 1988, twenty N'Dama calves were born at ILRAD's headquarters, making a total of twenty-four calves that have been produced by the embryo transfer techniques from the five 1984 heifers brought from the Gambia as frozen embryos in 1983 (ILRAD, 1988). Until the technology for freezing embyros was available, Kenyan quarantine regulations prohibited the importation of cattle from West Africa. Under experimental conditions in Kenya, these transplanted N'Dama have shown substantial resistance to tsetse-transmitted *T. congolense* infections compared to Boran calves reared in the same environment. Frozen embryo technology has created an opportunity to test the hypothesis more fully that trypanotolerant livestock breeds can make a valuable contribution to increasing meat and dairy production in central and southern Africa (Vermeer, 1986).

Recent developments in molecular biology and biochemistry, such as monoclonal antibodies, chromosome mapping, and DNA probes, have enabled ILRAD scientists to develop methods to diagnose different strains (serodemes) of each trypanosome species. Improved techniques suitable for diagnosis of the disease in small laboratories or in the field should be available in the near future (ILRAD, 1986b). Identification of different parasite strains can help in recommending drug treatments for infected livestock and will be helpful in vaccine research.

Training became a major component of the Trypanotolerant Livestock Network in 1982, a year after the project's inception. As the network has expanded both in area and research scope, demand has grown for technicians and scientists equipped with the appropriate skills. To help fill this need for skilled personnel, ILCA, ILRAD, and ICIPE jointly operate two annual seven-week training courses, one in English and the other in French (ILCA, 1986b). Initially, students spent a month at ILRAD learning about livestock health and diseases, two weeks at ICIPE, where they concentrated on the life cycles of the twenty-two species of tsetse flies known to transmit trypanosomiasis, and one week at ILCA studying livestock productivity. For logistical reasons, training has recently been concentrated at ILRAD's Nairobi facilities. The content and duration of the course remain the same. In 1983, ILCA, ILRAD, and ICIPE published a training manual for the network that also contains a sample of scientific literature on trypanosomiasis (Murray et al., 1983).

The Trypanotolerant Livestock Network has increased our understanding of trypanosomiasis, but many details about the disease are still unclear. Trypanotolerance can be associated with the ability to limit quantities of the parasite in the blood, the ability to control the duration and frequency of parasitaemic waves, and the capacity to prevent or resist parasite-induced anemia. These tolerance traits can be expressed singly or in combination (ILCA, 1986b). The gene systems responsible for tolerance have yet to be identified. More studies of the physiologic responses underlying the ability to control trypanosome parasites and associated anemia are needed to understand how genes responsible for tolerance function. The Trypanotolerant Livestock Network is expected to play a key role in expanding our knowledge of many of these puzzling aspects of the disease.

Potato Research Networks: PRECODEPA and SAPPRAD

Five major multipurpose commodity networks involving the potato have been assembled in Latin America, Africa, and Asia (Appendix 1). All of them were established, and in some cases are still coordinated, by CIP.

In each of CIP's regional networks, research responsibilities are divided up among participants according to who is best equipped to tackle the problem (CIP, 1984a). In the case of the Southeast Asian Potato Program for Research and Development, which was established in 1980, the research pie is divided accordingly: tropical agronomy (Indonesia), simple seed production (Papua New Guinea), germplasm development (Philippines), true potato seed (Sri Lanka), and rustic storage of seed potato (Thailand). In 1984, a further joint project was added to SAPPRAD: technology transfer. The rationale behind this addition is that in many cases promising technologies are already available, but they need more field testing and promotion.

The first regional potato research network to be set up was PRECODEPA in 1978. A Rockefeller Foundation scientist, John Niederhauser, was the instigator and early guiding light for this ambitious collaborative research network, which now spans ten countries in Central America and the Caribbean. The Swiss Development Cooperation recognized PRECODEPA's potential and quickly provided funds for meetings and training. CIP administers the network's external funding and provides technical advice and training assistance as needed. PRECODEPA soon took hold because potato research programs in the region were generally small and needed help in tackling the many obstacles to increasing potato production and consumption.

Like its sister network in Southeast Asia, PRECODEPA participants have divided up tasks according to the national programs in the region most able to handle the particular topic (Table 14). Leadership of particular projects is by election of the participants, rather than mandated from above. PRECODEPA, now ten years old, has matured into a relatively smooth, self-managed operation. Neither CIP nor the SDC has to exert a hands-on management style to keep the network going.

An organizational structure that encourages active participation in

Table 14. PRECODEPA projects, their leaders, and the participating countries, 1988

Project	Leader	Coleader	Participants
Seed production	Mexico	Cuba	Costa Rica, Dominican Republic, Guatemala, Nicaragua, Panama
Postharvest technology	Guatemala	El Salvador	Honduras
Bacterial diseases	Costa Rica	Haiti	Guatemala, Mexico
Nematodes	Panama	Mexico	Costa Rica
Potatoes for humid, lowland tropics	Cuba	El Salvador	Dominican Republic
Tuber moth	Costa Rica	Guatemala, Mexico	Dominican Republic
Late blight	Mexico	None	Costa Rica, Cuba, Dominican Republic, El Salvador, Guatemala, Panama
Germplasm enhancement	Mexico	None	Costa Rica, Guatemala, Panama

management by network members and clearly defines the duties of the coordinator greatly facilitates the routine functioning of PRECODEPA (Table 15). The management flow chart of PRECODEPA contains three main entities: the coordinator, the regional permanent committee (COPERE), and the executive committee (COE). The position of coordinator has been fully funded since 1987, allowing greater flexibility and freedom in attending to network matters. The coordinator is elected for a two-year term and is responsible for implementing decisions reached by COPERE and COE.

PRECODEPA is really run by its policy committee, COPERE, which sets policy and establishes research priorities. COPERE, composed of two representatives from each of the ten participating countries and

Table 15. Governance structure of PRECODEPA and the roles of each management and policy entity

Regional Permanent Committee
(COPERE—Comité Permanente Regional)

Functions
Nominate executive committee (COE)
Draw up regional projects for potato improvement
Establish priorities and appoint project leaders
Seek external funds for individual projects
Help national programs set their potato research budgets
Contract an international auditor for reviewing expenditures of external funds
Set up a technical committee
Composition
Two representatives from each participating country and two representatives from CIP

Executive Committee
(COE—Comité Ejecutivo)

Functions
Implement COPERE's decisions
Contract PRECODEPA personnel
Promote PRECODEPA's linkages with other networks and technology transfer
Organize annual PRECODEPA meetings and other technical reunions
Represent PRECODEPA when dealing formally with countries and institutions
Composition
President, a technical adviser, and an administrative secretary

Coordinator

Functions
Arrange annual meetings
Act as liaison between technical and administrative aspects of project implementation
Coordinate technical reviews
Manage external funds
Maintain regular contact with external donors
Carry out agreements reached by COPERE and COE

two CIP scientists, meets once a year to review projects and explore new research directions. COPERE has the power to terminate projects that are not going well or have fulfilled their task and to initiate fresh research projects.

COPERE does not need prodding from external donors or institutions to keep projects on track. If a lead country is unable for any reason to carry out its role in a project, it may be dropped from the project or assigned as a participant rather than as the leader in that project. Such policy and management shifts are undertaken by members without apparent rancor because all players realize that the credibility and viability of the entire network suffer if a member is not performing satisfactorily. Members of COPERE therefore do not constantly look for inputs from the SDC or CIP; they keep their own house in order.

Far from feeling anxious about their diminished role as PRECODEPA matures, the SDC and CIP are encouraged by the network's progress both scientifically and managerially. They are pleased that they do not have to hire consultants or send members of their own staff to solve problems. Both the donor and the implementing agency are always ready to provide advice and assistance, but only when asked. This nonintrusive role of the external donors and technical advisers is particularly evident at the annual meetings, at which helpful suggestions are offered rather than passed down as policy recommendations or even instructions.

PRECODEPA places strong emphasis on training and monitoring tours as a means of strengthening the network. As of 1988, some 150 technicians and scientists from member countries had attended seminars, workshops, and short courses organized by PRECODEPA. These training and educational exercises are usually conducted in member countries, but CIP also hosts some courses. As a means to bolster further the exchange of new research methodologies, CIP invited all potato program leaders in PRECODEPA countries to its headquarters in Lima, Peru, for a week in November 1988 to witness firsthand laboratory and fieldwork related to potato improvement. Two types of monitoring tours are conducted: coordinator's field visits and trips organized as an adjunct to annual meetings. In the first case, the coordinator, sometimes in conjunction with one or more technical assistants, attempts to visit the potato program of each PRECODEPA country at least once a year. Field trips to research stations and on-farm trials are organized for all participants in the annual network meeting and offer an opportunity for many of the scientists to see each others' projects since the venue of the annual meeting varies. In 1988, for example, the PRECODEPA annual meeting was held in Guatemala; the 1989 meeting took place in El Salvador.

Apart from the stimulation provided by the accelerated flow of ideas and information, collaborative research networks focusing on potato have begun to yield some tangible benefits. Not surprisingly, the oldest of these potato networks, PRECODEPA, has made several valuable contributions to potato research in Central America and the Caribbean. One outgrowth of Mexico's participation in PRECODEPA is the development of potato varieties with varying degrees of resistance to late blight; one such variety, Tollocan, with moderate resistance to the pathogen, was released in Mexico in 1981 and has also been adopted in Panama, Costa Rica, Honduras, and Guatemala (CIP, 1985a). By systematically sharing information and technology, participants in PRECODEPA gain access to good ideas and productive and stable potato varieties.

In addition to the dissemination of potato varieties, PRECODEPA has helped spread the use of tissue culture for rapid multiplication of desirable potato clones, improved rustic storage techniques, and led to better seed production (CIP, 1985a). Six PRECODEPA countries now have seed improvement schemes as a result of Mexico's technical capability and leadership. Rustic storage methods for seed or consumer potatoes have been improved by constructing simple huts made from local materials that allow diffuse-light to penetrate. A major advantage of diffuse-light storage for seed potatoes is that sprouts are in better condition and emerge faster than those kept in the dark. Plants derived from tubers kept under diffused light mature faster, thereby escaping some pest damage. Diffuse-light storage huts for seed potatoes are used particularly in Guatemala and Costa Rica and to a lesser extent in Panama. In Guatemala, farmers traditionally store seed potatoes in boxes inside their already cramped houses; separate diffuse-light sheds for seed potatoes liberate domestic space and can be used for storing other items after planting.

Farmers in Panama, Honduras, and the Dominican Republic are also increasingly acquiring this intermediate technology for seed potato storage. In the Dominican Republic, a rice deficit in 1983 spurred the government to provide generous incentives to potato farmers. The result was overwhelming; more potatoes were harvested than could be stored properly. Instead of opting to build expensive cold storage facilities, technicians from the Ministry of Agriculture were dispatched to Guatemala, the nodal country in the PRECODEPA network for rustic storage research, to learn more about this technology. After

Figure 16. Diffuse-light storage for seed potatoes in a farmer's backyard. Diffuse light inhibits sprout elongation and improves vigor. Note funnel and plastic bag baited with pheromone to trap tuber moths under eaves. Near Fang, northern Thailand, June 1986.

discussing various designs and ideas in Guatemala, the technicians returned to the Dominican Republic, where the government authorized the building of five large storage facilities for consumer potatoes using sugarcane stalks and pine poles (CIP, 1984b). These diffuse-light storage facilities were ready a month after the problem was identified, and the potatoes were stored for six months with less than 2 percent losses. Diffuse-light storage for potatoes is also reaching Asia through a sister network, SAPPRAD. The Thailand national program is testing this storage technique near its Fang research station in the northern highlands close to the border with Burma (Figure 16).

Both PRECODEPA and SAPPRAD are increasingly concerned about assisting the process of technology transfer in addition to their primary function of research. Much more scientific investigation is needed on several fronts in potato work, particularly in locating and employing genetic resistance to pests and pathogens, but interesting and high-quality products are not very useful if they do not reach farmers. PRECODEPA and SAPPRAD scientists are therefore exploring ways to form closer working relationships with extension services in the countries embraced by their respective networks.

CIP has fostered a series of collaborative research networks focusing on potato because many developing countries cannot afford a comprehensive potato research program. A strategy of pooling resources is especially useful to small countries with few scientists and limited research facilities (CIP, 1984b). This strategy has paid dividends with PRECODEPA and other multipurpose potato networks. For example, PRAPAC, a CIP-instigated potato research network embracing Burundi, Rwanda, Uganda, and Zaire, has strengthened the farming systems dimension to research on potato and other commodities in Rwanda (Fresco and Poats, 1986).

Collaborative Research Support Programs (CRSPs) for Peanuts and Bean/Cowpea

The Peanut Collaborative Research Support Program, one of eight CRSPs that have been sponsored by the Board for International Food and Agricultural Development[1] and USAID, is designed to help alleviate constraints on the production and fuller use of peanuts for cash and subsistence. The Peanut CRSP unites scientists at twenty-two institutions in fourteen countries in Africa, Southeast Asia, the Caribbean, and the United States.

Specific projects in this multipurpose commodity network and main locations where the research is focused include breeding for resistance to foliar and soil-borne diseases (Senegal, Burkina Faso, Niger, Texas A&M University); mycotoxin management in peanut by preventing contamination (Senegal, Texas A&M University); integrated pest management (Burkina Faso, University of Georgia); optimum use (Caribbean, Sudan, Alabama A&M University, University of Florida); varietal improvement (Caribbean, Philippines, North Carolina State

1. Additional information on the history and scope of CRSPs can be found in Chapter 1.

University, University of Georgia); arthropod management (Philippines, Thailand, North Carolina State University); biological nitrogen fixation (Philippines, Thailand, North Carolina State University); mycorrhizae (Philippines, Thailand, Texas A&M University); and consumption and storage (Philippines, Thailand, University of Georgia).

The Peanut CRSP has benefited the United States and collaborating countries in the Third World in several ways. Researchers at participating U.S. universities and in Nigeria recently discovered that peanut mottle disease, also known as rosette disease, is caused by a pair of viruses. Although the disease is not present in North America, the viruses might spread there. Fortunately, scientists at ICRISAT and the University of Georgia have developed a modified enzyme-linked method to detect the pathogens which is proving especially helpful to countries that have no specialist plant virology laboratories (ICRISAT, 1986).

Peanut CRSP activities have also helped disseminate GH 119-20, a peanut variety developed by B. B. Higgins at Experiment, Georgia, to Senegal.[2] Although GH 119-20 was never widely adopted in the peanut-growing areas of the United States in the Southeast and Texas, it has caught on in Senegal and is spreading to other parts of West Africa.

The Bean/Cowpea CRSP has helped farmers in California and several developing countries. This network facilitated the discovery of heat-tolerant cowpea germplasm, which is being used to develop varieties of black-eyed peas, as the crop is known in the United States, that thrive in hot weather.

Soil Management Networks

Among the many factors influencing crop yields, inherent soil fertility and fertilization practices, including the application of organic mulches, are often pivotal. An understanding of soils and their dynamics is essential to devise agricultural systems that can sustain high yields over a long period. Identifying deficiencies of trace elements in soils, achieving a proper balance of nutrients in fertilizer applications, learning when and where to apply fertilizer or mulches, and learning

2. *Peanut Collaborative Research Support Program (CRSP): A Three Year Summary 1982–1984*, Peanut CRSP, University of Georgia, Georgia Experiment Station, Experiment, Georgia/United States Agency for International Development, Department of State, Washington, D.C. (Grant No. DAN-4048-G-SS-2065-00).

how to manage soils for crop production can mean the difference between anemic yields or fruitful harvests. Not surprisingly, several collaborative research efforts are under way to study these questions for different farming environments.

Joint planning and standardization of methodology appear to be major concerns of international soil fertility and management networks. Indeed, the desire to generate data that can be extrapolated to wide areas and avoid using outdated sampling and testing procedures is central to the mission of soil fertility networks. Although international soil fertility networks are not numerous, interest in them is growing and new ones are being planned. The need to increase and sustain yields in a finite world with a rapidly growing population is fanning the widespread interest in soil fertility networks in developing countries.

To increase the efficiency of fertilizer uptake by rice and thereby reduce costs to farmers, IRRI and the International Fertilizer Development Center in Muscle Shoals, Alabama, coordinate the International Network on Soil Fertility and Sustainable Rice Farming. INSURF links more than sixty scientists in twenty-one nations in Asia, Africa, Latin America, and the Caribbean (Appendix 1). IFDC sends the participants various fertilizer formulations to test free of charge; recipients must cover the cost of applications, harvesting, and data recording. IRRI coordinates data collection from participating sites, analyzes the information, and reports the results. Trials comparing organic mulches, particularly using water ferns, with inorganic fertilizers are being conducted at IRRI as part of the institute's contribution to the network. IRRI also provides INSURF's coordinator and advice from economists, who have assisted the network by devising a standardized questionnaire for interviewing farmers on fertilizer use.

After thirteen years of operation as of 1989, INSURF has begun to bear fruit. One useful result from fertilizer trials is the discovery that deep placement of nitrogenous fertilizer (urea supergranule) near the roots of rice plants, combined with basal application of slow-release, sulfur-coated urea, results in better yields than surface dressing with prilled urea (Mamaril, 1985; INSFFER, 1986). In some cases rice yields have increased by as much as one ton per hectare by using the twin application method. This yield advantage is particularly evident with moderate to low levels of nitrogenous fertilizer application (less than 87 kg/ha) and when percolation is not pronounced (Mamaril, 1984).

Another network focusing on soil fertility and management at widely scattered sites around the world was the Benchmark Soils Project (BSP), which the University of Hawaii launched in 1974 with support from USAID. This network, which phased out in 1982, linked up to twenty-five sites in five countries and was established to gauge how experience gained from cropping on one soil family could be transferred to similar but widely separated soils (Beinroth et al., 1980; BSP, 1981, 1982a, b). BSP explored the feasibility of transferring soil management practices associated with maize, sorghum, rice, and root crops among three soil families: Typic paleudults (sites in Cameroon, Indonesia, Philippines), Hydric dystrandepts (Indonesia, Philippines, Hawaii), and Tropeptic eutrustox (Hawaii, Brazil, and Puerto Rico).

One of the goals of the Benchmark Soils network was to break down isolation among soil scientists and reduce the number of redundant trials. Literature in the agricultural sciences is replete with fertilizer trials, often conducted at considerable expense, but with little discussion about the soil type, growing environment, and the significance of the results to the broader picture. Other objectives of the network were to help tropical countries evaluate the potential of upland areas for intensive cropping and to demonstrate the value of soil classification for agricultural development.

Three major findings resulted from BSP experiments. First, crops responded similarly to phosphorus fertilizer across widely scattered sites within the same soil family. Second, maize gave the highest yields in warm, dry Tropeptic eutrustox soils and the poorest harvests in Hydric dystrandepts. Third, application rates for phosphorus fertilizer to achieve a given yield were similar in Typic paleudults and Tropeptic eutrustox soils; additional phosphorus was needed to achieve the same yield in Hydric dystrandepts.

By 1982, the initial phase was completed and the network evolved into the International Benchmark Soils Network for Agrotechnology Transfer. IBSNAT spans thirty benchmark sites in twenty-six countries and builds on BSP's eight years of experience testing the feasibility of transferring agrotechnologies within the same soil family based on the U.S. Soil Taxonomy, a powerful classification system. Like its predecessor, IBSNAT is funded by USAID.

IBSNAT's three main goals are to accelerate the flow of agrotechnologies from sites of origin to new locations, maximize the successes and minimize the failures in agrotechnology transfer in the tropics and subtropics, and assess the long-term effects of agricultural practices on soils. IBSNAT worked initially on four cereals (maize, rice, sorghum, and wheat), three grain legumes (kidney bean, groundnut, and soybean) and several root crops (aroids, cassava, and potato). The list of crops within IBSNAT's research plan has grown as collaborators request them and sufficient interest is expressed by other participants.

For each of the selected crops, IBSNAT organizes a computerized data base containing information on fertilizer trials, soil, weather, and crop management. This data base is used to model cropping systems under different soil management strategies. When perfected, the simulation models should enable scientists to predict such aspects as daily crop growth, soil erosion, and the impact of different planting dates. The versatile models will be able to predict the effects of changing only one component, such as fertilizer rate, or a combination of management practices. In this manner, it is hoped that at least some ideas can be tried out on a computer before expensive field experiments are laid out.

As of 1985, the CERES crop model for wheat and maize had been tested globally by IBSNAT and was found to predict crop performance with reasonable accuracy (IBSNAT, 1985). The SOYGRO crop model to simulate growth, development, and yield of soybean is being widely tested. Models for the other crops within IBSNAT's program are under various stages of development. The data base management system was fully operational in 1989.

The network's headquarters in Honolulu administers the overall program and assembles the data base, which can be tapped by all collaborators (IBSNAT, 1985). The network is highly decentralized. USAID supports IBSNAT activities through subcontracts to some of the collaborators, including the University of Puerto Rico at Mayaguez; the Soil and Water Research Laboratory operated by the Agricultural Research Service (ARS) at Temple, Texas; the Institut National de la Recherche Agronomique (INRA), Toulouse, France; and the University of Florida. USAID has also funded collaboration between IBSNAT and the Soil Management Support Services (SMSS), the Soil Conservation Service (SCS), National Soil Survey Laboratory (NSSL), IFDC, and Michigan State University. IBSNAT's headquarters facilitates rather than controls the network's activities by organizing monitoring tours, workshops, and symposia.

The coordinator oversees the network with the assistance of two advisory committees and a management committee (Table 16). The Technical Advisory Committee (TAC), established in 1984 with five leading international scientists, provides advice on such matters as systems analysis and crop models. The Collaborators' Advisory Panel (CAP) was set up in 1985 with four members from national, regional,

Table 16. Governance structure of IBSNAT

Technical Advisory Committee (TAC)
Functions
Update project on most recent advances in information and computer technology as they relate to agricultural development
Advise project on design criteria for decision support system for agrotechnology transfer (DSSAT)
Advise project on potential collaborators to network
Annually review program achievements and make recommendations for any changes
Composition
An international committee selected to represent the full range of agricultural disciplines; the six members include a systems scientist/ecologist, a crop modeler/agronomist, a soil scientist, a plant breeder, a social scientist/economist, and a crop protection specialist
Collaborator's Advisory Panel (CAP)
Functions
Assess project outputs from the standpoint of user groups
Evaluate outputs and recommend changes
Composition
Three members, including a director of a national agricultural research program, a director of a regional agricultural research program, and a staff member of an international agricultural research organization
Management Committee
Functions
Review project management and administration including performance of principal investigator and host institution
Review fiscal management and establish budgetary priorities based on recommendations of TAC and CAP
Composition
Principal investigator, a project monitor from donor agency, and an administrator from the host institution

and international organizations. CAP's purpose is to seek ways of applying IBSNAT's research to agricultural planning and development. The Executive Management Committee (EMC) was established in 1984 and is composed of five members drawn from collaborating institutions and the donor agency and is responsible for advising the coordinator on management and project implementation issues. The network is thus endowed with a sound management and technical input system and is poised to provide tangible benefits for agricultural development in the tropics.

Members in developing countries are enthusiastic about the network, and many are participating at their own expense. A sense of ownership of the network coupled with ready access to sizable international data bases concerning crop production and crop management on major soils may explain the high degree of commitment in developing countries to IBSNAT.

Outlook

Collaborative research networks are likely to grow in number and kind because many decision makers have become convinced that international collaborative efforts in agricultural research can be very rewarding to all participants. Such networks are often envisaged as the answer to complex or widespread problems that cannot be handled adequately by unlinked national efforts. Despite the growing interest, however, collaborative research networks will require more careful planning and management and can result in sacrificing some individual independence in research to a planned group effort.

10

Networks as International Centers

A new organizational form to carry out agricultural research has emerged recently in the international arena that is considered by some as a hybrid between networks and international agricultural research centers (IARCs). Networks as international agricultural research centers (NIARCs) combine benefits of networking with the institutional advantages of international centers. NIARCs were formed in the 1970s and particularly in the 1980s to further research on agroforestry, bananas and plantain, aquaculture, and management of freshwater and coastal fisheries. Other NIARCs are on the horizon to work on vegetables and aquaculture. To understand better the origin and development of NIARCs, we first explore their history.

International Centers

One of the great success stories in agriculture this century has been the establishment of the international agricultural research centers. The concept was pioneered by the Rockefeller and Ford foundations with the establishment of the International Rice Research Institute (IRRI) in 1960. Since then international centers have been key players in agricultural research for the tropics and subtropics and have helped improve the effectiveness of national programs in developing countries.

The international centers have several characteristics that distinguish them from other agricultural research organizations. Each is an autonomous organization formed to improve agricultural research in developing countries, especially the tropics and subtropics. The international centers are conceived as workplaces for critical masses of world-class scientists focused on major crop or animal commodities or topics needing special services and policy analysis.

International centers are governed by boards of trustees composed of distinguished individuals from both industrial nations and developing countries. Board members do not represent any organization or country; each serves in his or her individual capacity. A center's board selects the director general, who assembles staff to work with the board and center leadership in carrying out the center's mandate, which is assigned by the CGIAR or a center's governing board. The centers draw scientists from around the world.

Thirteen of the international centers are supported by the CGIAR, which is cosponsored by FAO, UNDP, and the World Bank (Baum, 1986; Plucknett and Smith, 1982). More than forty donors, including bilateral aid agencies, multilateral development organizations, private foundations, and governments in developing countries, support the CGIAR. The World Bank provides a chairman and secretariat for the CGIAR in Washington, D.C. The CGIAR is served by an independent Technical Advisory Committee, consisting of a chair and fourteen members, half of whom are citizens of developing nations.

The payoffs emanating from international centers working in partnership with national programs have spawned international agricultural research centers operating outside the CGIAR. New international centers have been established for important crops or problems, such as the Asian Vegetable Research and Development Center in Taiwan and the International Fertilizer Development Center (IFDC) in Alabama. These non-CGIAR centers emulate the principles and concepts adopted by the early CGIAR centers.

IRRI, the first IARC, was established near the College of Agriculture of the University of the Philippines near Los Baños on fertile land ceded by the Philippine government. IRRI was provided with new facilities especially equipped for crop improvement and related research. Since IRRI was established two decades ago, a large share of its research has been conducted at headquarters. Relationships with national programs are forged through training, outposting of IRRI staff, and cooperation through networks and other collaborative mechanisms.

Specific crops or commodities have benefited greatly from the international center approach. International centers help create a greater awareness of research needs with specific commodities and other aspects of agricultural production. Working relationships with national programs in developing countries and, increasingly, laboratories in industrial nations have been tested and improved. International centers have learned new, effective ways to facilitate the research of national programs in the Third World.

Not all international problems can be handled satisfactorily by the usual model of an international center, however. It is hard to conceive of an international center for each of the hundreds of domesticated food, fiber, beverage, and medicinal plants. Some international problems are characterized by numerous constraints, most of which require site-specific investigation. In some cases, institutional capacities to work on a specific problem are so meager and dispersed that a single research center model would be ineffective. In such cases, the agricultural research community has begun to design international centers that work mostly as networks or that use networks as their main operational arms. NIARCs have thus formed to fill a niche in the global agricultural research system.

Networks as International Agricultural Research Centers (NIARCs)

NIARCs present some interesting contrasts to the international center model (Table 17). NIARCs have the strength of an independent research institution with an independent board of trustees and a direc-

Table 17. Contrasting features of networks, networks as international agricultural research centers (NIARCs), and international agricultural research centers (IARCs)

Feature	Networks	NIARCs	IARCs
Legal identity	Usually not	Yes	Yes
Can receive and administer funds	Usually not	Yes	Yes
Can employ staff directly	Usually not	Yes	Yes
Assigned mandate	No	Yes	Yes
Board of trustees	No	Yes	Yes
Independent governance	Usually not	Yes	Yes
Owns its headquarters	No	Seldom	Yes
Owns its substations	No	Seldom	Usually
Research capability at headquarters	Sometimes	Sometimes	Usually
Strategic planning	Steering committee/coordinator	Board/management	Board/management

tor as the chief executive. A NIARC has an international mandate assigned by its donors; the mandate and a program strategy arrived at by an independent board help improve funding prospects.

A NIARC is a more structured and formal entity than a network. Because of its institutional identity, a NIARC can receive, administer, and disperse funds on its own, rather than through a third party as in the case of most networks. A NIARC also employs its own staff and does not have to borrow personnel from other organizations.

A NIARC is a focal point for identifying priority research topics as well as addressing major research needs. A NIARC monitors the quality of its own research as well as the performance of its cooperators. A NIARC can make long-range plans and ensure proper training and information systems because it has a long-term institutional capacity. Furthermore, it can build up its own library, information services, data processing, and other research support facilities. Data bases and information exchange are of particular importance to NIARCs. Other organizational advantages include matching research needs with scientists in national programs, carrying out in-house research to meet special needs, and mobilizing funds for its core activities as well as for programs of its collaborators.

A NIARC is also unusual in that it organizes its own networks and yet may participate in networks coordinated by other institutions. Planning of network activities is also different because the board of trustees plays an important policy role in establishing the long-term direction and strategy of the NIARC and any network it supports.

National research goals and aims of a NIARC may at times conflict. The historic tension between the flow of funds to international centers, as opposed to national programs, also surfaces with NIARCs. Some argue that funds for international agricultural research would be more profitably channeled to national programs than to international centers. Still, the total budget of the CGIAR system is close to what Brazil alone spends on agricultural research. NIARCs are hardly likely to divert significant funding from national programs.

Nevertheless, the potential for conflict exists and could be serious because most NIARCs rely on networks to achieve their goals. The key players in networks are national programs, and difficulties between NIARCs and their clients can hamper joint research efforts.

Another potential problem for NIARCs is that the coordinators for their networks are employees of a particular NIARC. Part of the program goals are set by the NIARC, and the coordinator is a NIARC employee, so some national program scientists may not feel that they own the network.

An inherent weakness of most network-centers is that they are usually formed as a mechanism to strengthen research in an area in which expertise is thin. Thus a NIARC is seen as a way to unite scattered and often inadequately trained individuals. In such situations, building enthusiasm for collaborative efforts may be difficult, and national programs may see international efforts as competing with their own work in an already underfunded field.

Each network-center needs to assess how much in-house research is warranted and to what degree it should rely on networking. Another issue is how much research should be undertaken at headquarters. Some NIARCs began operations believing that no in-house research capability was needed. It appears that some NIARCs have had second thoughts about this approach after discovering that networks can be thin technically and devoid of new ideas to energize the research. The conflict-of-interest issue again arises. In cases when a NIARC plans to do some research at headquarters as well as at selected locations in regional or thematic networks, in-house NIARC research may compete with networks for funds and other resources.

Some NIARCs will now be profiled to illustrate how they operate and address some of these issues. Several characteristics, such as ownership and control of facilities and the degree to which networking is employed, vary considerably between NIARCs (Table 18). Like networks and international centers, NIARCs occupy a relatively wide and varied niche along the broad spectrum of organizations involved in international agricultural research.

International Network for the Improvement of Banana and Plantain (INIBAP)

The International Network for the Improvement of Banana and Plantain was established in 1984 to coordinate an international research effort on bananas and plantains in developing countries. Head-

Table 18. A comparison of characteristics of some existing networks as international agricultural research centers (NIARCs)

NIARC	Facilities		In-house research	Headquarters research	Library	Thematic networks	Regional networks
	Owns	Rents					
INIBAP	No	Yes	Some	Some	No	No	Yes
IBSRAM	No	Yes	No	No	No	Yes	Yes
ICLARM	No	Yes	Yes	Yes	Small	Yes	No
ICRAF	Yes	Yes	Yes	Yes	Small	No	Yes

quartered in Montpellier, France, INIBAP is an autonomous, non-profit organization under international law and enjoys privileges and immunities granted to other international institutions by the French government. INIBAP's raison d'être is to increase the productivity and stability of banana and plantain production by smallholders. The main clients of this NIARC are national programs and small farmers.

INIBAP operates from a small headquarters and depends on regional networks to further the work program (Chart 5). Research is conducted primarily by network members. A small staff of scientists at headquarters gives leadership in specialized fields and helps provide scientific leadership to the regional networks.

INIBAP has a donor support group, a board of trustees, and a director. The board of trustees has eleven members, one ex officio host country nominee, the director (also ex officio), five members from banana- and plantain-producing countries, and four members ap-

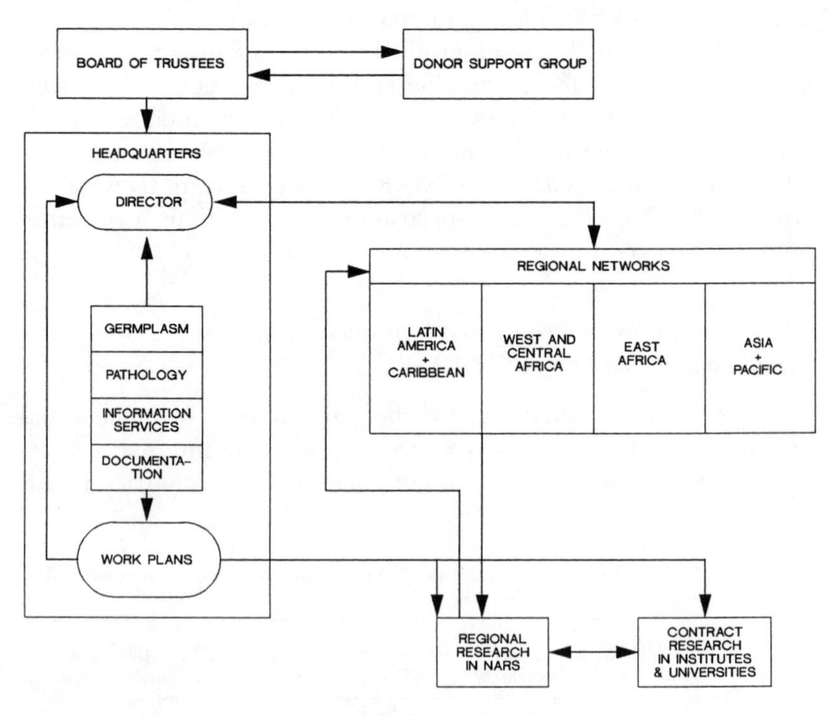

Chart 5. Organization of INIBAP showing relations between headquarters, regional networks, national agricultural research programs (NARSs), and other outside research institutions

pointed on the basis of their scientific or managerial skills. After reviewing and approving strategies and work plans, the board of trustees recommends a work program and budget to the donor support group. The board also periodically reviews the networks.

The regional networks focus mainly on germplasm development, pathology, and information and documentation exchange and receive technical support from headquarters. Each regional network is coordinated by a full-time INIBAP scientist.

The principles under which INIBAP operates include decentralization, close relations with national agricultural research systems, thematic research on pantropical problems, regional research on key problems, a coordinating role for headquarters, and dependence on linkages with other institutions in both developed and industrial countries. These connections are made via networks or contract research. Interregional coordination is the director's responsibility.

As of May 1989, three regional networks had been established covering Latin America and the Caribbean, West and Central Africa, and East Africa. A fourth regional network is planned for Asia and the Pacific (Chart 5). Regional networks are established following initial workshops or fact-finding missions at which founding documents are written (INIBAP, 1986, 1987; Jaramillo, 1988). The regional networks are coordinated by scientists chosen on the basis of their high qualifications and interpersonal management skills.

Priorities for regional networks are exchanging information and germplasm and detecting research needs. Regional networks are expected to strengthen national banana research programs by organizing workshops and other collaborative activities, facilitating regional and global germplasm trials, sharing information, providing small research grants, helping national programs elaborate funding proposals, and providing training for national program staff. Differentiation in the networks is encouraged through country specialization and division of labor based on national needs, capabilities, and interests.

Progress in thematic research is considered essential to establish a secure footing for the network. This concept echoes the principle that a network must have a strong knowledge base to drive the collaborative effort. INIBAP considers it necessary to have specialists at headquarters to oversee thematic research worldwide, rather than depend on the regional coordinators to carry out all the work. Regional coordinators participate in thematic research within their respective geographic areas but are not responsible for the progress of the work worldwide. INIBAP's research strategy is based on the premise that regional coordinators are unable to develop research with enough scope to address global needs so thematic research responsibilities are also included at INIBAP's headquarters.

Funding needs for INIBAP are envisioned in three categories: core funding for INIBAP at headquarters, funding for network activities, and bilateral funding for national banana research programs. INIBAP uses its core funds to launch studies of germplasm and diseases.

INIBAP has no facilities of its own. Its headquarters facility is provided by a French government agency. Regional offices are located at collaborating institutions. Like a network, then, INIBAP does not own any facilities. INIBAP strives to catalyze and strengthen research at the national level by forging interorganizational linkages.

International Board for Soils Research and Management (IBSRAM)

The International Board for Soils Research and Management grew out of an awareness by soil scientists and others of the need for an international initiative to harness reservoirs of expertise in soils research and to focus soil management research to meet better the challenge of boosting food production. IBSRAM is conceived as a coordinating center to work with national programs and international agricultural research centers. IBSRAM is one of the few NIARCs to list networking in its statement of objectives. The networking approach is seen as a way to validate existing knowledge, promote applied soil management research by national programs, foster training, and provide technical support.

The board of trustees consists of eleven members, four of whom are from developing countries. The board of trustees acts much like the board of an international agricultural research center in that it hires the director, decides on the program priorities and strategies, and approves the program and budget for submission to donors.

A 1983 workshop in Townsville, Australia, identified four priority areas for soil management research: wetland soils, vertisols, acid tropical soils, and land clearing and development. Specialized international workshops were then held on each of these topics. Based on deliberations at these workshops, IBSRAM decided to establish three regional networks. Two IBSRAM networks operate in Africa, the Network on Management of Vertisols under Semi-Arid Conditions in Africa (MOVUSAC) and the Network on Land Development and Management of Acid Soils in Africa (AFRICALAND), while ASIALAND covers Asia and the Pacific. IBSRAM was incorporated in Australia in

1983 but moved to Bangkok, Thailand, where it has not yet gained recognition as a legal international organization.

IBSRAM considers its clients to be national program scientists in developing countries. Its networks are seen as bridges between international centers and national agricultural research systems. IBSRAM's networks provide a decentralized way to promote adaptive research on soil and land management in developing countries. The regional networks help ignite enthusiasm and incorporate national researchers into a larger research effort.

IBSRAM's networks divide research on the basis of national interest, capabilities, and needs. Participating national programs draw up project proposals for the work to be done by national scientists. Each network has a network coordinating committee that reviews national project plans and submits them to IBSRAM's board of trustees for approval. Because these plans are completed and submitted on an individual country basis, some network activities are slow to get started.

IBSRAM does not do research itself. Research is conducted by the national program scientists. IBSRAM staff members are mostly coordinators and facilitators for the networks. Research plans for IBSRAM and its collaborators are made by committees and consultants. IBSRAM considers that strategic or basic research needs of the networks can be handled through collaboration with advanced institutions in industrial nations.

In its network activities, IBSRAM acts mostly as an implementing agency for bilateral donors. IBSRAM receives and disburses funds in the network to national programs in accordance with donors' wishes and network plans. IBSRAM networks involve common core experiments agreed upon at inception seminars. At such meetings, satellite experiments at national sites are planned. Baseline studies are used to choose one or two experimental sites in each country.

International Center for Living Aquatic Resources Management (ICLARM)

The International Center for Living Aquatic Resources Management was established in the Philippines in 1977 as an autonomous, nongovernmental Philippine corporation. This NIARC was established to stimulate and strengthen fisheries and aquaculture research in Asia and islands in the Pacific. ICLARM has recently extended its reach to Africa and Latin America.

ICLARM sees its role as complementing and supporting national and regional institutions dealing with fisheries, aquaculture, and resource management of coastal zones. To accomplish these broad goals, ICLARM has acquired a small core staff of competent and highly productive scientists at headquarters in Manila who complement research by field staff posted at national institutions. ICLARM rents office space for its headquarters and shares national or regional research facilities. ICLARM has only one field station of its own, in the Solomon Islands.

ICLARM has always emphasized interdisciplinary research. Its strategy has been to document problems carefully before taking on a new initiative. Another important part of its strategy is to emphasize publication of research efforts, and these published documents have played an important role in research planning with the center's national program collaborators.

A major characteristic of ICLARM's program has been a strong commitment to networking and decentralized research. The center sees networking as a way to help national programs improve their capabilities. Such networks usually begin with information exchange and the building of individual and institutional capacities, but their ultimate purpose is scientific consultation or more intensive collaborative research.

By 1989, ICLARM had launched four networks: the Network of Tropical Aquaculture Scientists, the Network of Tropical Fisheries Scientists, the Asian Fisheries Social Science Research Network, and the Coastal Aquaculture Network. Also, the Coastal Area Management Project with six Asian countries has many features of a network, including its organization, planning, and operational style.

ICLARM was formed to conduct research and to assist national research programs. Because ICLARM has virtually no facilities of its own, the center conducts training on an individual basis through collaborative research or on a group basis through short in-country or regional courses organized in collaboration with other institutions.

ICLARM is governed by a fifteen-member board of trustees. Two board members are ex officio. The board, as the policy-making body, appoints the director general, approves the program and budgets, and reviews the program and management of the center.

Although ICLARM is most like an international center of the four NIARCs reviewed here, networking is its central mode of operation. Although ICLARM may be too small to be a scientific powerhouse in its own right, it plays an important energizing and leadership role. Network meetings provide periodic contact with clients for information exchange and joint planning. High quality scientific publications and newsletters strengthen communication with national program scientists.

International Council for Research in Agroforestry (ICRAF)

The International Council for Research in Agroforestry was established in 1977 to administer a comprehensive program leading to better land use in the tropics. In 1978, the council moved its headquarters to Nairobi, Kenya. After leasing office space in downtown Nairobi for several years, ICRAF moved to its own facility on the outskirts of the city. In 1981, a new strategy was adopted to focus more efforts on acquiring in-house capability to understand and analyze land-use systems and to identify agroforestry technologies. Particular emphasis was placed on overcoming diagnosed constraints and problems in land-use systems.

To accomplish this reorientation, ICRAF collaborates closely with national programs, but it faces problems such as the difficulty of defining agroforestry, the poor knowledge base, and the lack of a niche for agroforestry within cooperating organizations. Because of its global mandate, ICRAF seeks to develop widely applicable methods for strategic, applied, and basic research from its headquarters and a field station at Machakos, Kenya.

ICRAF's strategy and program are set by a twelve-member board of trustees. The strategy in working with national programs is to identify national institutions as potential partners, and to establish a national steering committee and a national task force to work with ICRAF scientists. A memorandum of understanding is then drawn up between ICRAF and the country concerned. Next comes an analysis of existing production systems, planning and design workshops, and training courses. From these national projects and efforts, ICRAF then establishes regional or zonal networks to share information and conduct research on shared problems.

ICRAF has decided to focus its networks on Sub-Saharan Africa through its Agroforestry Research Networks for Africa (AFRENA). Four AFRENA networks have been established, the Unimodal (rainfall) Upland Plateau of Southern Africa, the Bimodal Highlands of Eastern and Central Africa, the Humid Lowlands of West Africa, and the Semi-Arid Lowlands of West Africa. Networks are also proposed to train and provide information to scientists in South Asia, Southeast

Asia, and Latin America. ICRAF is considering the establishment of subregional offices to assist its research networks.

NIARCs on the Horizon

The CGIAR has studied several possible new initiatives that are potential candidates for the NIARC model. Each is considered too complex or having too many site-specific characteristics to be tailored for an IARC model. Some of the ideas under consideration for two of these proposed NIARC initiatives will be discussed to highlight some of the constraints and factors considered.

Aquaculture

Subsistence and commercial fish culture is a relatively new activity in most developing countries. Some expertise in aquaculture research exists in a few places, but most programs are new, staff are inexperienced, and many research problems have yet to be tackled. Areas of research envisioned for international support include genetics and nutrition of pond fish.

A new aquaculture NIARC would be independent, self-governing, and autonomous. The network-center would have a board of trustees to develop policy and grant research contracts. A program advisory committee would assist the board.

An aquaculture NIARC would consist of a central research/coordinating group, probably located in Asia. The NIARC would embrace two international networks conducting research on the nutrition and genetics of pond fish; units in Africa and Latin America to assist technology transfer and coordination; national, regional, and international research institutions conducting contract research on priority topics; and a group of universities in developed countries to provide technical assistance.

The major concern of the small scientific core group at headquarters would be research, not development. The core group would also help stimulate and serve the networks through scientific leadership.

Vegetables

Development organizations and donors have been interested for some time in improving research on vegetables in developing countries. The Asian Vegetable Research and Development Center was established with this purpose in mind, but because of its location in Taiwan, AVRDC has experienced some difficulties operating in certain areas. The CGIAR has agreed that an organization is needed to foster research on vegetables, especially tropical vegetables, and to encourage the transfer of technology among developing countries. The problem is difficult because more than one hundred vegetables are grown in a wide variety of ecological and cultural conditions in over one hundred developing countries.

The NIARC model was chosen for vegetable research so that a central coordinating body could be formed to develop collaborative research networks. Coordination among global networks focusing on specific vegetables, or groups of closely related vegetables, would be improved under a NIARC umbrella. Most of the research would be carried out by national scientists working in their home institutions under contracts with the coordinating body. Their research would be augmented by research at headquarters and by contract research with advanced facilities. AVRDC is envisioned as a major participant in a global research enterprise on vegetables. Crops to receive priority, at least initially, include tomatoes, peppers, onions, okra, eggplant, and leafy vegetables.

Outlook

Interest in compelling research problems that need international research attention on a continuing basis almost guarantees that further efforts will be made to establish networks that operate much like international agricultural research centers. In most of these cases the NIARC model will be chosen because the usual IARC model will be considered inappropriate or difficult to adopt.

The question of headquarters research capacity has been a major subject of debate. Some persons argue that headquarters research efforts may dilute or downplay networks. Others argue that networks would be weak and that without strong research capability at headquarters, few new ideas or methodologies would infuse research efforts. It is generally agreed that headquarters research need not take precedence over network needs and that an effective review process could help ensure this. The coordinating body will carry out much of its research in collaboration with national programs.

11

Problems and Remedies

In some circles, a perception is emerging that networks may not be living up to their expectations. A spreading sense of concern about networks spurred an international conference in 1989 to examine the track record of networks in development research and training.[1] In Africa, particularly, some scientists and donors are increasingly skeptical whether networks promoted by the Consultative Group on International Agricultural Research and various agencies of the French government are helping agricultural research on the continent (Eicher, 1988). Others argue that collaborative partnerships between international centers and national programs need to be emphasized in Africa and that the selective use of networks is one way to achieve that objective (Oram, 1988).

Controversy surrounding the network approach to agricultural research stems from the often glaring gaps between a network's potential and its actual performance. Like any other organization, networks are prone to difficulties, and an analysis of their problems can help bolster their effectiveness. Here we attempt to identify some of the constraints to realizing the potential of networks.

Networks are proliferating and evidently benefit research in agriculture and other scientific fields, but even the most successful ones encounter problems. Indeed, the spectacular progress of some of the

1. The Asian and Pacific Development Center (APDC) and the Commonwealth Secretariat sponsored an international forum on Network Experience in Development Research and Training at the APDC, Kuala Lumpur, Malaysia, 29 June–1 July 1989.

more dynamic networks may attract so many adherents and add on so many projects that research programs become unwieldy. Some of the problems explored here can be attributed to departures from tenets for effective management of networks established in Chapter 3 and successful networking principles outlined in Chapter 5. Not all of the difficulties examined here are unique to networking; many also apply to individual research efforts.

Our review of problems is not intended to discourage the formation of new networks. Rather, we wish to bring difficulties to the attention of research administrators and scientists so that pitfalls can be avoided with new networks and corrective action taken with existing ones. An examination of problems associated with networking and possible remedial measures will benefit current and future networks.

Problems encountered in agricultural networks can be grouped into three broad categories. The first category relates to the conduct and quality of research, particularly methodological issues, uneven feedback of results from collaborators, data management, priorities and scope of the research agenda, planning, and proper characterization of study sites. The second major group of networking difficulties involves personnel issues such as rapid turnover of participants, the paying of collaborators, and language barriers. The third category is essentially institutional and includes such aspects as disbursing arrangements for funds, accounting difficulties, the potential of networks to distort national programs, and inadequate credit and extension services to convey technology to farmers.

Research Quality

Quality of data or other products is a major concern in some material exchange, scientific consultation, and collaborative research networks, particularly those dealing with international nurseries and agricultural machinery. Two components of research quality are analyzed here: acquiring and processing sufficient data to make research progress and reach reasonable decisions, and the quality of the information.

Inadequate communication between network participants and coordinators is a common complaint even in successful networks. Occasional dissatisfaction about the quality of communications has arisen in material exchange, scientific consultation, and collaborative research networks.

Communication difficulties result from lack of motivation on the part of participants and unreliable mail service. Many national programs in developing countries must use international telephone calls, courier services, and telex sparingly because of budget constraints. Mail delivery in many Third World countries is sporadic and untrustworthy. Courier firms are more reliable but are expensive. Some network sites have no telecommunication or mail services.

PRECODEPA, the Pond Dynamics Collaborative Research Support Program, and the Bean/Cowpea CRSP are among the networks in which poor communication has sometimes stirred apprehension. PRECODEPA overcame communication problems with the La Esperanza Experiment Station in Honduras by using the offices of the Swiss Development Cooperation in Tegucigalpa, the capital of Honduras, to channel PRECODEPA mail and messages for Honduran scientists doing potato research.

Inefficient processing of information can retard the progress of a network. Information management is essential for effective networking. In the past, the Asian Rice Farming Systems Research Network was plagued by the lack of an efficient computerized data management system (Carangal, 1988). At the eighteenth meeting of the Working Group of ARFSN held in Pakistan in 1987, agreement was reached among collaborators to iron out differences in the way data emanating from networking trials are processed.

The falling price of computers and the increased use of satellites for telecommunication are improving data processing and opening up electronic mail services worldwide. Electronic mail allows instant communication between participants at a fraction of the cost of air couriers, telephone calls, or telexes.

Some networks suffer from poor feedback, particularly during their early stages. In the case of the seventh International Pearl Millet Adaptation Trial (IPMAT) in 1981, for example, seed packets were sent to forty-seven locations in fourteen countries, but the coordinating body in India, ICRISAT, received results from only twenty-five locations in eight countries (ICRISAT, 1984). CIMMYT, coordinator of the International Bread Wheat Screening Nursery, was obtaining feedback from only 36 percent of the sites participating in the global network (Dubin and Rajaram, 1982).

Low returns from international nurseries are caused by a variety of resource and program difficulties. On the human resource side, weak motivation of participants stands out as particularly relevant. Excuses for not performing one's part range from lack of fertilizers to insufficient land to plant the trials. Yet scientists will always find room, funds, or personnel for a project that interests them and is likely to yield tangible benefits. Participants in international nurseries and other networks are more motivated if they see clear payoffs from collaboration.

Motivation to participate fully in international nurseries would be keener if fewer trials were burdened with lines that have not been properly checked out in advance. National scientists sometimes complain that too much "junk" germplasm is fed into international nurseries. Improperly tested materials sap the research effort by unnecessarily tying up limited personnel, equipment, supplies, and land. More prescreening of materials would reduce the amount of poor yielding or vulnerable lines incorporated into international nurseries. IRRI, coordinator for the International Network on Genetic Enhancement of Rice, prescreens rice lines for rainfed conditions by orchestrating observational and yield nurseries in the Philippines, Northeast Thailand, and Bangladesh. Observational trials involve several hundred entries, and the best performers qualify for yield trials. Only the best lines in yield trials are submitted to INGER.

Another way to improve the proportion of nursery sites reporting results to the coordinator is to issue preliminary reports of early returns. By alerting participants about outstanding performers before all results are tallied and analyzed, collaborators have time to request superior lines for the next test cycle. To speed up reporting of results and entice fuller cooperation, ICRISAT began using a computer in 1983 to issue preliminary results of IPMAT trials.

Even when collaborators in an international nursery report back results, the data may not always be reliable. The INGER coordinator carefully checks all data returns from nursery sites. Tests are performed on the returns, such as checking the coefficients of variability, to detect impossible yield figures or unlikely scores on susceptibility to diseases and pests. In 1975, only 30 percent of INGER returns were considered reliable; by 1985, some three-quarters of the nursery results were deemed accurate (D. V. Seshu, pers. comm.).

Indifferent quality also plagues agricultural machinery networks. Many participants in such networks are small-scale workshops, often not much bigger than a village blacksmith shop, and quality of the finished product varies markedly. In some parts of the Third World, small-scale operators have served farmers badly by selling them shoddy machines and implements (Bell et al., 1985). Large-scale manufacturers certainly do not have a monopoly on quality, but volume producers of agricultural machinery and implements usually have more experience and built-in quality control procedures. Large-scale manufacturers often eschew agricultural machinery networks because they cannot market exclusively products based on blueprints drawn up by public institutions.

Observant engineering students are quite capable of discerning improperly assembled agricultural machinery. In 1986, one of the authors accompanied a field trip for IRRI trainees from several countries to small- and medium-scale agricultural machinery workshops near Manila and Los Baños in the Philippines. Some of the students on the field trip whispered criticisms of the poor workmanship witnessed in several shops.

In Asia and increasingly in other regions, Japanese firms or their subsidiaries dominate the market for many agricultural machines such as power tillers, for small-, medium-, and large-scale farmers. Large manufacturers benefit from economies of scale so their products are highly competitive. Agricultural machinery networks may need to identify their niche a little better and concentrate on locating manufacturers able to produce high-quality products that will build a client base.

Quality is not just the concern of networks dealing with germplasm or machinery. Networks whose primary product is research information rather than biological materials or implements can also suffer from uneven research output from participants. Farming systems research in eastern and southern Africa still leaves much to be desired in spite of sustained training efforts by CIMMYT staff (CIMMYT, 1985). The indifferent quality of some farming systems research in the region has hampered progress of the CIMMYT Eastern and Southern Africa Economics Program (CIMMYT/ESA; Appendix 1). The Southeast Asian Universities Agroecosystem Network (Appendix 1), another farming systems network with a strong emphasis on resource management, has also had problems with deficient data collection and sloppy analysis (Rambo and Sajise, 1985).

Whenever possible, networks adopt a common, or at least very similar, methodology. The methodological approach does not have to be a straitjacket that all must fit; some leeway should always be allowed for scientists to adapt research to local conditions. But if participants stray too far from a commonly accepted way of conducting the joint research effort, results may not be comparable.

Examples of research difficulties arising from using different methodologies can be drawn from a variety of networks. In the case of the

International Network on Soil Fertility and Sustainable Rice Farming, IRRI took on the responsibility of analyzing some soil samples because several national programs used different techniques for assessing the chemical and physical properties of soil (Mamaril, 1985). National program scientists use different techniques for analyzing soil samples because they do not have access to the same equipment or supplies; some scientists thus resort to older and less precise methods or are unable to run certain tests.

Divergent methodologies have also stymied farming systems research in some instances. Disparate efforts in farming systems networks reflect the lack of a clear consensus on what constitutes farming systems research rather than any inherent organizational problems with the networks themselves. In some cases, farming systems research is still in definitional and methodological flux. Indeed, there are almost as many definitions of farming systems research as practitioners of it (Francis, 1986). What now passes for farming systems research has traditionally been conducted under a wide variety of labels such as cropping systems or human ecology research.

The questions addressed by farming systems research are heavily influenced by the disciplinary background of the investigators. Farming systems research has been, and continues to be, dominated by agricultural economists, agronomists, and anthropologists. The problem is that such disciplines have rarely worked as a team. Thus the flavor of farming systems research tilts heavily in favor of the dominant, or sole, discipline in charge of the research. CIMMYT's farming systems research is dominated by economists, whereas in Thailand, agronomists conduct much of the farming systems research.

Effective farming systems research requires inputs from many disciplines. Communication problems between disciplines still need to be ironed out, but there is a growing realization that social and hard sciences have much to learn from each other. When mutual respect is evident, multidisciplinary teams can work together. The evolution of farming systems research will probably lead to more of a team approach involving a mix of scientists from various disciplinary traditions. The exact blend of scientific specialties will doubtless vary by crop and farming conditions, but it is vital to keep an open mind and retain a holistic approach. Then as bottlenecks are identified, the appropriate skill areas can be drawn into the research effort.

Differing interpretations of farming systems research have prevented a fuller integration of the social sciences into agricultural research and development, particularly in some Third World countries.

The Malaysian Agricultural Research and Development Institute (MARDI), for example, eschews the ARFSN because of its perceived lack of focus and its preoccupation with the small farmer. Senior administrators and scientists at MARDI feel that small farmers will always play a role in Malaysian agriculture, but much of the food for urban centers and agricultural exports will come from medium- to large-scale operations.

An exclusive emphasis on small farmers may not always be in the interests of national agricultural programs. Equity concerns are not necessarily best served by focusing most research and funding on the smallest farmers. Increasing food production and agricultural exports rest on the efforts of small-, medium-, and large-scale farmers. As much as possible, farming systems and other networks should therefore focus on scale-neutral technologies.

As stated in Chapter 5, networks need to define clearly the problem or set of problems to be addressed and draw up a realistic research agenda. The most successful networks run the greatest risk of departing from these two principles because their high profile and achievements inevitably attract more followers and can lead to overly ambitious research agendas. As a network grows, new collaborators often lobby to add research projects. The balance between flexibility and disciplined adherence to an established research program is delicate. Networks are expected to evolve, take on new challenges, and follow fresh lines of inquiry. But if more and more tasks are taken on without others being completed or dropped, the enterprise can become overloaded and stall. Not only are managerial problems more likely with a larger network, but the research thrust can become diffuse.

The Asian Rice Farming Systems Network is one example of a successful international cooperative effort that may have crossed the gray zone between excessive conservatism and taking on too many tasks. ARFSN's mission has expanded dramatically since its inception in 1975 and now includes projects such as the integration of livestock with rice cropping, rice's niche in various crop rotations, the performance of rice under different fertilizer regimens, and the impact of weeds on rice yields. ARFSN is considering research into such areas as the role of women in rice farming and fuller integration of aquaculture with rice cultivation. These are worthwhile research topics, but ARFSN may not be the optimal framework to house all of them.

Material exchange and collaborative research networks can become overwhelmed by too many activities. The Micro-Biological Resources

Centers (MIRCENs)[2] may have strayed from their original focus on microorganisms useful in nitrogen fixation and industrial fermentation processes. To help reduce farmers' dependence on pesticides, some MIRCENs have recently branched out into biocontrol research. Some of the agents currently used in biological control of crop pests are microorganisms, such as certain bacteria and fungi, but many predators and parasites of crop pests are insects and other arthropods. It remains to be seen whether MIRCENs' involvement in biocontrol research bears fruit, or whether a separate biocontrol network is called for.

One solution to the danger of listening to siren songs in collaborative research is to encourage the establishment of separate networks to tackle items that do not match, or at least complement, the original research agenda. Here we run into the dilemma of whether to lump or split when planning networks. Packaging sometimes disparate topics into a single research enterprise is tidier for administrators but may not make sense from the research viewpoint. Splitting some of the larger networks into smaller subnetworks, or creating new networks instead of joining existing team research programs, may be more sensible in some cases. Carried too far, though, networks could multiply exponentially. The Special Program for African Agricultural Research is trying to rationalize networks in West Africa, starting with those serving maize and cassava, and perhaps going on to soils networks.

Yosuf Hashim, a researcher who has come up through the ranks to assume the role of director general of Malaysia's agricultural research system (MARDI), believes that networks are best kept lean and sharply focused. When pondering MARDI's involvement in exchange or collaborative research networks, the director general generally prefers small-scale networks set up to tackle specific problems that can readily be disbanded once the task has been accomplished.

An often overlooked issue in the conduct of research in networks is the proper characterization of study sites. Unless the areas where data are being gathered are properly described, results may not be comparable and extrapolations are risky. Accurate characterization of research locations is particularly vital to international nursery work and soil networks.

Improper identification of environmental parameters at research sites has been a problem in some soil fertility and germplasm testing

2. The history and purpose of MIRCENs are discussed in Chapter 1.

networks (Greenland et al., 1987). Careless surveys of the chemical, physical, and climatic conditions of sites can lead to soil management recommendations that do not hold up in other areas thought to be similar or to crop yields that do not match results obtained from other sites in the nursery. One of the functions of the coordinator of IBSNAT is to assist collaborating countries in properly identifying soil sites within the network.

Soil identification is a burr under the saddle for quite a few networks. Several soil classification systems are still in use worldwide, ranging from FAO/UNESCO's system to the USDA Soil Taxonomy. Rough equivalents can usually be found in each system, but precise analogs are difficult to establish. The USDA Soil Taxonomy is the finest grained classification system and has the additional merit of using a scientific binomial nomenclature that follows clearly defined subdivisions. For these reasons, it has been the most widely used soil classification system since the early 1970s. But it has drawbacks. First, Soil Taxonomy was devised in temperate U.S. climates and does not always work well in tropical environments (Whitmore, 1985). Second, the names are complex and difficult to remember. Similar criticisms were launched at Linnaeus when he proposed the now standard scientific nomenclature for plants and animals in the eighteenth century. Resistance to Soil Taxonomy, admittedly a tongue-twister for many soils, will surely decline as more scientists become familiar with the terminology.

Personnel Problems

Rapid turnover of participants is high on the list of complaints of those involved in networks. Constantly changing staff in farming systems research in Kenya, for example, has retarded training courses conducted by the CIMMYT/ESA farming systems program (CIMMYT, 1985). Concern has been expressed that expected promotions of staff working in the Philippines on the Soil Microbiological Barriers project of the Peanut CRSP will adversely affect that portion of the network's program (Cummins, n.d.). Rapid turnover of potato scientists in Honduras, triggered by changes in political leadership at the national level, have hampered some projects within the ten-nation PRECODEPA network. A similar problem has been noted with Papua New Guinea's participation in SAPPRAD, a sister regional potato network operating in Southeast Asia. Frequent changes in staff in

Ecuador's bean research program have hampered Bean/Cowpea CRSP activities there. SPAAR is working to address the turnover problem in African networks.

Uncertainties and delays caused by excessive staff turnover are not confined to national agricultural research programs in the Third World. Scientists from developed countries on temporary assignment to bilateral or larger networking activities must eventually return to university or other institutional duty, usually after only a few months or a year of participation in the project. Just as they become familiar with the project's scope and issues, they must often return home. The Bean/Cowpea CRSP, the Small Ruminants CRSP, the Soil Management CRSP, and the Sorghum/Millet CRSP, among other networks, have been beset by frequent changes in personnel contracted by USAID, the main sponsor of CRSPs. A replacement may not always be found immediately. Such lack of research continuity on a problem is self-defeating (Borlaug, 1986).

When a collaborator leaves a network, a delay usually ensues until a new member comes on board. A replacement may be delegated, in which case he or she may not be enthusiastic about the network. Even when the replacement decides to join voluntarily, the research effort inevitably slows while the new person becomes familiarized with procedures. Committee work is usually hindered when a new member joins the group and has to learn developments and plans discussed at previous meetings.

Rapid turnover of participants is especially critical during the early phases of a network. One or two individuals often provide the catalyst for establishing a network; if they leave too soon, the joint effort may dissipate. Personnel changes can be accommodated more easily once a collaborative research program has matured.

At any point in the life of a network, however, the departure of a key individual can disrupt the joint effort. In smaller networks, key individuals play an especially prominent role, such as the coordinator, by virtue of their dynamism, diplomatic skills, and keen scientific insights. Ideally, all participants in a network have equal responsibilities and inputs, and network operations should be as routine as possible and not hinge on the constant intervention of any one individual. In practice, though, a few "stars" often provide the leadership and many of the insights. When a prominent member of a network leaves, other participants may not be able to take up the slack effectively. A major reason that PRECODEPA (Appendix 1) dropped its socioeconomics project in 1988 is that Guatemala, the anchor for this emerging concern for farming systems research within the regional potato network, no longer had sufficient staff trained in the relevant disciplines after the promotion or departure of some key individuals.

In Third World countries, the best scientists are often tapped early in their careers to occupy administrative posts. Networks therefore often lose an appreciable number of members to promotion. Once promoted to executive positions, scientists are usually cut off from the front lines of research and rarely resume their scientific careers. Networks are a mechanism for research and usually involve institutional arrangements with various nodes, but each node consists of people to whom the webs, spokes, or lines that convey information are attached. The ultimate success of a network rests largely on the quality of the individuals involved.

Better training facilities and more rewarding careers in the agricultural sciences are crucial if the quality of participants in agricultural research networks in the Third World is to improve. Several developing countries have devoted considerable resources to improving the technical capabilities of their agricultural research staff. Mexico, for example, devotes U.S.$2.6 million to postgraduate education for agricultural scientists; currently, some 10 percent of the Mexican agricultural research program's 2,200 staff members have doctoral degrees, and the proportion is growing. Brazil and India have also invested heavily in postgraduate training for their agricultural research scientists, and this sensible policy has enabled those countries to participate more fully in international networks and has helped boost and sustain crop and livestock yields.

Language differences sometimes pose problems in international networks. English is the international language of science and commerce, but not all participants in networks are fluent in English. After some instruction, many persons are able to read a foreign language, but they find it more difficult to understand conversation or to speak the language. Network participants can usually read newsletters and instructions in English, but how much they derive from monitoring tours and workshops depends largely on their command of English. Simultaneous translation is expensive and not usually available at network meetings.

Two major languages are spoken in some networks, and this can lead to some confusion in such matters as the use of terminology. In the case of the West African Farming Systems Research Network, for example, English and French are the two most widely spoken languages in the region. No single African language has currency over large parts of the region, as does Swahili in East Africa. Francophone countries in West Africa have no desire to adopt English because of custom and strong scientific ties with France.

The schism between Anglophone and French-speaking countries in West Africa is a perennial problem, but it is not insurmountable. A program to combat river blindness in West Africa has effectively spanned both Anglophone and Francophone countries (Walsh, 1986a). A climate of mutual respect, concentrated effort to agree on methodology and terminology, and determination to solve problems can usually overcome communication barriers posed by more than one language being spoken in a network.

Institutional and Bureaucratic Problems

Poor planning plagues some networks, particularly in their early stages. Obstacles crop up in some cases because the network is informal and no government-level agreements have been made to expedite visas and the shipment of equipment and supplies. Whenever possible, networks should remain informal, but as more countries join international material exchange and collaborative research networks, the need to formalize agreements between governments will increase.

Network coordinators and steering committees also need to pay closer attention to other planning matters such as better coordination and timing of monitoring tours and training courses. The efficiency of some networks with multiple projects would be improved if there was better coordination among the subnetworks or facets of the network (Greenland et al., 1987). Sufficient lead time for all network activities is essential for a smooth operation. Sometimes instructors for training courses are given short notice on when they are expected to give classes or conduct field trips. In 1983, for example, the Nairobi-based International Center of Insect Physiology and Ecology was given only three weeks warning to prepare a course for trainees in the ILCA-coordinated Trypanotolerant Livestock Network.

Funding arrangements pose periodic problems in even the most successful and productive networks. Two major facets of the funding problem stand out: insufficient funds to accomplish the task—a perennial complaint by scientists the world over—and the uneven flow of funds. The question of insufficient funds can usually be resolved with proper planning, a clear goal and realistic research agenda, and, most important, some early successes to entice donors. In an age of tightening funds for research, credibility becomes an ever more important asset.

The problem of oscillating funding levels needs careful attention by network coordinators and their advisory boards. A research program can be inundated with funds at one stage, then struggle through a prolonged financial drought. Research demands on staff time and resources also follow a pattern of peaks and troughs, yet funding disbursements are not always in rhythm with the varying pace of scientific work. Insufficient funds forced the coordinator of the African Research Network on Agricultural Byproducts (Appendix 1) to suspend publishing the network's newsletter in 1982. Fortunately, the funding picture brightened in 1983, enabling ILCA to resume publication of ARNAB's newsletter.

Sometimes shifts in funding procedures or withdrawal of support by external donors can undercut network efforts. In response to changes in USAID funding policies in September 1985, for example, the IRRI Industrial Extension Network now restricts its activities to the Philippines. From 1980 to September 1985, USAID, the major donor to IIEN, gave annual block grants to the network's coordinator to facilitate networking among six participating nations. After September 1985, funding for the network could be obtained only at the discretion of USAID offices in participating countries. The increased administrative load and funding uncertainties of such a move triggered the decision of IIEN's coordinator to drop the international portion of the network in late 1985. And the funding pullback of several major donors to UNESCO in recent years has constrained the growth of MIRCENs (Appendix 1). A change in UNESCO's leadership in 1987 might signal a return to healthier financial times for that United Nations agency and an upsurge in the worldwide activities of MIRCENs.

The flow of funds for networks is sporadic partly because most material exchange and collaborative research networks receive external funding from several donors, each with different schedules for releasing grants. The coordinator of the Trypanotolerant Livestock Network has worked out an informal agreement with the budget officer at ILCA, the network's coordinating institution, so that money is available to bridge funding gaps. Budget directors are essentially extending loans in such instances, and they are usually prepared to do so only when the network is performing well and external donors are reliable. International funding agencies normally deliver on promises and commitments when the enterprise they are funding is viable. Flexibility in the accounting divisions of institutions coordinating networks and high-quality performance by network participants can ease the difficulties created by the tidal nature of scientific funding.

A major reason why international agricultural research centers often coordinate international networks is that they are usually considered fiscally sounder than national programs. ILCA, for example, with a 1985 budget of $15.7 million provided by twenty-six donors, had earned an unsecured $1 million credit line with a commercial bank (Hardin et al., 1986). Many international agricultural research centers have similar flexibility to overdraw their accounts temporarily to prevent the disruption of research activities.

Another way to even out the funding flows from donors to networks would be to create some framework for coordinating support. At the moment, most donors, including multilateral and bilateral agencies, act independently. A "broker" organization could be set up for each large international network or group of networks to pool funds from external donors and release them according to need.

Although this idea has some merit, two difficulties surface immediately. First, donors may feel that they lack adequate control over their contributions. Glory for the success of a network may go to the scientists, research institutions, and broker organization, leaving external donors out of the limelight. Second, another level of bureaucracy between scientists and donors could be counterproductive. Bureaucracies absorb funds and have a tendency to grow ever larger and more complex. The administrative overhead of a broker organization would inevitably siphon away some money earmarked for laboratories, libraries, and the field.

Even when outside assistance in underwriting some of the costs of participating in networks is reliable, prolonged and generous financial support without appreciable contributions from participants can be counterproductive. Inadequate support from national programs has been identified as the weakest link in most scientific consultation and collaborative research networks in Africa (SPAAR, 1986). The West Africa Rice Development Association, an international agricultural research center based in Bouake, Côte d'Ivoire, has paid member countries in the region to plant rice nurseries. Returns from such heavily subsidized nurseries have been extremely disappointing; not only have numerous sites not reported data, but the information has often proved unreliable. Another problem with paying institutions to participate in networks is that collaboration usually ceases as soon as the funding ends.

In addition to difficulties related to outside funding, network scientists are vulnerable to unpredictable support from their home institutions. Agricultural research budgets in general have come under severe pressure with the global economic downturn in the early 1980s. Oil-exporting Third World countries have been especially hard hit by the sharp dip in the world price of oil. In Indonesia, oil exports account for 70 percent of foreign exchange and more than half of government tax revenues.[3] Under the strain of lower oil revenues, Indonesia slashed its operating budget for agricultural research by 50 percent in 1985. Such drastic cuts inevitably lead to the elimination of some projects and the scaling down of others. In 1986, budget cuts forced the Bogor Institute for Food Crops (BORIF) to withdraw from participation in INSURF even though fertilizers were supplied free.

Budget shortfalls often result in delayed maintenance and postponed purchases of vehicles used by national agricultural research programs. Gasoline and diesel oil shortages and strict quotas on fuel use by government vehicles are also chronic obstacles to sustained field research in many developing countries. Lack of transportation for researchers, technicians, and equipment hampers networks, particularly those involved in farming systems research. Farming systems research is often a relatively new component to national programs, and vehicles may not be assigned for such work. Unreliable transportation has hampered fieldwork by farming systems researchers participating in the CIMMYT Eastern and Southern Africa Economics Program and the West African Farming Systems Research Network (CIMMYT, 1985).

Vagaries in the flow of funds are only one dimension of the financial problems of networks. Sometimes the accounting of how those funds have been spent is inadequate, thereby jeopardizing the network. Misuse of funds is less of an issue than differences in bookkeeping methods and failure to produce accounting statements on time. Networks require a great deal of paperwork, too much in the minds of some scientists. Nevertheless donors require a strict accounting of disbursed funds, or their support may be temporarily, or even permanently, cut off.

Reporting procedures for network expenditures can be complex. In the case of PRECODEPA, for example, the participants must contend with a multitier accounting procedure (Chart 6). The network's donor, Swiss Development Cooperation, receives reports on how PRECODEPA funds are spent from CIP, which distributes funds to

3. *Economist*, 2–8 May 1987, p. 36.

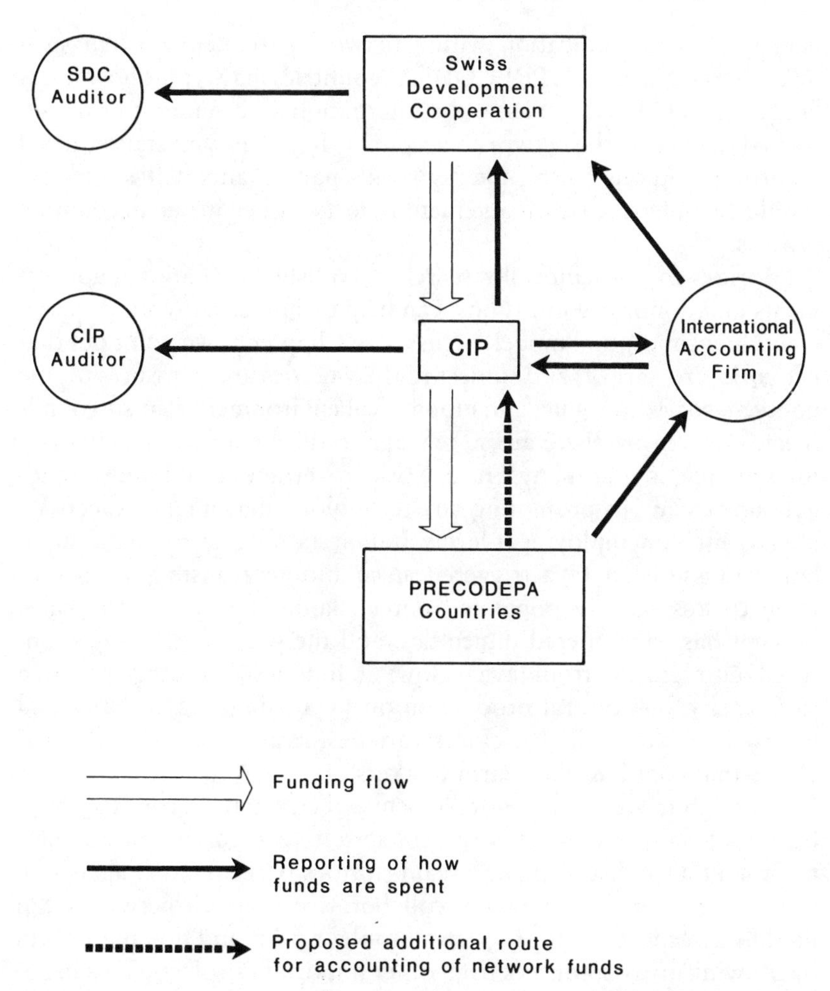

Chart 6. Funding flows and accounting procedures for PRECODEPA, a collaborative research network focusing on potato improvement in Central America and the Caribbean

PRECODEPA countries, and from an international accounting firm. These reports are then forwarded to SDC's auditor. If discrepancies are noted, policy guidelines dictate that funds be withheld until the situation is clarified. Some network accounting figures did not tally in 1987, apparently because of misunderstandings about how to report carryover sums from the previous year. This was a relatively simple problem to correct; it highlights, though, a previous issue discussed, inadequate communication within networks. To help prevent such difficulties in the future, PRECODEPA countries may report how funds are spent to CIP as well as to the international accounting firm contracted to oversee the network's accounts. Together with standardized reporting methods for all the network's participants, CIP's auditors should be able to catch inadequate reports earlier in the accounting process.

Priorities for agricultural research, established by national governments and donor organizations, can help or hinder network development. In a conceptual model discussed in Chapter 3, we point out that networks are part of, and subject to the vagaries and pressures of, the political, socioeconomic, and biophysical environment that surrounds them. One reason there are so few agricultural machinery networks, for example, is that many Third World governments are understandably uneasy about promoting any technology that might exacerbate already high unemployment levels. Indonesia is a case in point. Population pressure on Java triggered an ambitious transmigration program to less densely populated outer islands, but this safety-valve concept has encountered difficulties, and the pace of officially sponsored outmigration from Java is slowing. Indonesia's strategy has been to intensify agricultural production on Java using human labor and livestock as well as high-yielding varieties, rather than promote machines that could displace farm workers.

Although in some cases government policies may discourage, or at least not foster, certain networks, collaborative research ventures have the potential of distorting national priorities (Greenland et al. 1987). Unless sufficient care is taken, collaborative research networks can literally sweep national programs into their orbit and dominate them by drawing away resources from other important projects. This problem is more likely to develop if there is pressure to start the network from an external donor and if the national programs are relatively weak. Before a national program becomes involved in an international network, the relevance and impact of that network on the overall strategy and goals of the national program need to be assessed. If the principle that self-interest motivates network participants is adhered to, collaborative research undertakings are less likely to overwhelm and distract national programs in developing countries.

Inadequate credit and extension services are sometimes a particularly weak link between research and the creation of products that farmers can use. Insufficient credit and extension agents are certainly not problems unique to networking; they apply to many agricultural research projects. Credit and extension difficulties are most acute in much of Sub-Saharan Africa, but they can also be found in parts of Latin America and Asia. In South Sulawesi, Indonesia, for example, lack of credit by local banks has proved a major stumbling block in the dissemination of IRRI-designed hand tractors (Reddy, 1984). Such problems are beyond the purview of network participants and coordinators. Government decision makers need to be persuaded that improvements in credit and extension facilities will lead to a better return on the research dollar. Brazil is exemplary in this regard; the country has some fifteen thousand extension agents as well as thirty thousand salespersons for seed and fertilizer companies to reach farmers with new technological advances (Abelson and Rowe, 1987).

Inadequate extension seems to be a particular problem for agricultural machinery networks. Extension agents are often not well versed in the latest farm machines and implements or in their suitability under different agronomic practices. Extension agents in Third World countries are generally more familiar with recommended crop varieties and appropriate fertilizer treatments than with machines and implements that could help farmers boost production. South Korea, with its well-balanced and highly professional extension service, is exemplary in promoting machines and implements appropriate to the ecological and socioeconomic mosaic of its many farming communities.

In some parts of Asia, farmers are denied access to loans to acquire machinery because they do not hold title to land.[4] One way to circumvent this obstacle is to make the machinery itself security for the loan, an arrangement commonly known as chattel mortgage. In return for not having to put up land for collateral, the farmer must make a substantial down payment. Although this requirement discriminates against those who cannot afford a large initial payment, it ensures that those who do secure the equipment will make every effort to use it wisely because they have so much at stake. Chattel mortgage is a widespread practice in Thailand, where credit is provided by private distributors and machinery is repossessed if payments are in arrears (Rijk, 1985).

4. The requirement of banks that loan applicants have title to land as collateral is also a problem in Brazil and other parts of Latin America, where many farmers do not have legal title to the land they live on.

A Need for Realism

Most problems faced by networks are not insurmountable, but it is important to admit that they can occur. Networks can be very useful, but they are not a panacea. A clear understanding and application of principles for success can help overcome common pitfalls. Unsuccessful networks should be restructured and made to function better or they should be closed down. Continued support for networks from donors and participating countries requires well-planned programs and activities, a commitment to excellence, and good stewardship of resources.

12

Lessons and Future Directions

Networking in agricultural research has penetrated every continent and has spread to virtually all areas of the agricultural sciences. Five main benefits have spurred the proliferation of networks (Plucknett and Smith, 1987). First, networks have a comparative advantage in tackling problems of broad regional or international scope. Second, networks can increase research efficiency, a major asset in times of tight funding for research. Third, networks facilitate the development of improved methodologies. Fourth, in spite of some communication difficulties, networks generally enhance interactions among scientists. More frequent communication helps identify significant problems and fosters the development of new ideas, improved research methodologies, and appropriate technologies. Finally, scientific consultation and collaborative research networks upgrade the skills of national program scientists in the Third World through their specialized courses, monitoring tours, and workshops.

Scientists and administrators alike extol the virtues of networking to promote agricultural research and development (Jin-Hua, 1980). The French government fosters networks that help its bilateral assistance programs, particularly in Africa (France, 1986). Networking is seen as especially helpful to small, resource-poor Third World countries that cannot afford full-fledged agricultural research systems of their own (IRRI, 1988). By strengthening national programs in developing countries, networks help nations improve their agricultural production (TAC, 1986a). Stronger national programs in turn fortify networks, increasing rewards for all participants.

The phenomenal growth in networking is fueled by a wide range of benefits. Specific payoffs of some networks have been pinpointed in previous chapters, but here we wish to focus on generic benefits to networking that are clearly generating a wave of networks of all types in both industrial and Third World countries.

General Payoffs

Although numbers alone do not validate the concept, scientists in both industrial countries and developing nations are evidently deriving considerable benefits from collaborative research. Some specific payoffs of individual networks have been identified. Now we explore several generic benefits that appear to be intrinsic to networking in general and research networks in particular.

Trust and confidence that build up gradually as collaborative research efforts progress and mature are valued spinoffs from networking. Scientists and technicians participating in networks frequently remark on the collegial bonds that are established as a result of participation in networks, especially if they involve collaborative research. These professional bonds are based on mutual respect and are strengthened by periodic interaction.

Such unquantifiable dividends of networking also accrue to scientists working in nonagricultural fields such as public health. Albert Sabin, developer of the widely used live polio vaccine, reminisced on collaboration between the United States, Czechoslovakia, and the Soviet Union in testing the vaccine: "The most important lesson to me is the respect, confidence and mutual trust that develop during cooperative efforts in a struggle against a common enemy" (Sabin, 1986).

Scientists are also enthusiastic about networking because collaboration can lead to insights that would not be possible if the research was carried out in isolation. Indonesian collaboration in the IRRI-coordinated International Network on Soil Fertility and Sustainable Rice Farming has awakened national scientists to the need to pursue fresh lines of inquiry (Satari, 1986).

Networking provides a means for national programs to contribute actively to research and development rather than passively receive technology perfected elsewhere. This involvement builds the confidence and morale of national scientists and provides incentives for students to seek careers in solving development problems in their countries.

Participation in international research networks can help justify the continuation of research thrusts within national programs. In the case of PRECODEPA, for example, potato scientists in Mexico have successfully argued for continued support for their research activities linked to the regional potato network on the grounds that they have an international obligation that would be embarrassing for Mexico if they had to pull back from, or even drop, because of budget cuts. Agricultural scientists in developing countries thus envisage networks as one way to further their particular research interests.

Administrators and directors of research organizations are generally enthusiastic about networking because more research can be accomplished at less cost (Satari, 1986). Although some additional buildings and glass or screenhouses are occasionally needed to participate fully in collaborative research projects, networks generally rely on existing facilities rather than embarking on new construction (IDRC, 1986:15). Networking also trims research costs because less work is duplicated. The time-saving benefits of networking are especially noticeable in international nurseries, where other participants have done much of the development of entries. International nurseries probably save national programs at least three to four years of breeding effort in getting new materials to farmers.

Networks can save time, but a sustained effort is still essential to achieve tangible results. Networks are no quick substitute for the arduous process of coming up with a usable product, whether it be a new wheat variety, an improved rice thresher, or recommendations for controlling crop and livestock diseases. In the case of the Trypanotolerant Livestock Network, for example, much of the groundwork was accomplished by a UNEP- and FAO-commissioned study in 1977. Only since 1983 have a sufficient number of people been trained and logistics ironed out so that high-quality data are being generated for in-depth analysis.

Donors are receptive to the concept of networking because existing knowledge can be put to better use. Scientists constantly emphasize how much more needs to be learned, but senior administrators and donors sometimes point to the mountain of research results that are rarely if ever used. Which viewpoint is closer to the truth depends on the research topic and geographic area, but networks can be effective both at bridging knowledge gaps and sharing existing information.

Researchers in the Third World often emphasize the benefits of international collaboration. In Asia, leading rice scientists are particularly pleased with INGER. The world's largest germplasm exchange network has benefited numerous rice-growing countries, particularly Indonesia, Bangladesh, and Thailand (Siwi and Beachell, 1980; Zaman, 1980; Awakul, 1980).

The benefits of international collaboration also accrue to developed countries. Industrial nations gain a great deal by exchanging agricultural technology and information (Wennergren et al., 1986). Farmers and consumers in industrial and developing countries have benefited from international collaborative research. C. J. Rodrigues, a leading Portuguese scientist, stresses that combined efforts by international teams are essential for tackling important research problems and claims that international collaboration has been indispensable in the fight against coffee rust, one of the most serious diseases of that beverage crop (Rodrigues, 1977).

Donors and agricultural research institutions are devoting increasing resources to networking. The International Livestock Center for Africa in Addis Ababa is expected to boost its support for livestock networks from $0.9 million in 1985 (6.8 percent of the center's total budget) to $2.5 million in 1990 (10 percent of the expected budget), a 30 percent expansion of the budget devoted to networking (Hardin et al., 1986). In the past, the World Bank's involvement in agricultural development has been mainly in infrastructure improvement such as new irrigation projects and surfacing roads. More recently, the bank has shifted its emphasis to financing national research programs, especially in Africa (Yudelman, 1985). One component of the bank's support for national programs is improving linkages with international agricultural research centers. This policy shift is expected to generate more information and technology for use in development projects.

Networks on the Horizon

One indication of the growing interest in international collaboration in agricultural research is the vast array of new networks on the horizon covering a wide range of topics, ranging from agroforestry to blue-green algae and fungi. Focus shifts from the microscopic in some aspiring networks to whole organisms and even large ecosystems in other planned collaborative efforts. One theme underlies many of the proposed networks: the need to devise sustainable agricultural systems. Thus more international nurseries are in the offing, as well as collaborative efforts to improve soil fertility by nitrogen-fixing organisms and to arrest the worrisome decline in forestry resources in many parts of the Third World.

Agroforestry research networks are planned to help solve the mounting fuelwood crisis in Africa and to meet growing fodder and food needs for livestock and people there. With anticipated support from the World Bank, the U.S. Agency for International Development, and Canada's International Development Research Centre, the Agroforestry Research Networks for Africa will be composed of several subnetworks covering different regions. AFRENA, groundwork for which was laid in 1985, will concentrate on developing improved seed of some five to ten fast-growing tree species, organizing agroforestry technical groups, and arranging training courses (Spears, 1985). AFRENA's projected operating budget for the first five years is $4.6 million.

The major function of AFRENA Sahel will be to assemble a diverse germplasm collection of multipurpose trees suitable for the extensive, drought-prone region. AFRENA Southern Africa will focus on multipurpose trees for fodder and soil fertility maintenance. This subnetwork will be coordinated by the Nairobi-based ICRAF with the assistance of ILCA and ICRISAT, headquartered in Ethiopia and India respectively. AFRENA East Africa will be coordinated by ICRAF with the help of ILCA and IITA. IITA will co-coordinate AFRENA West Africa with ICRAF. This subnetwork will work closely with ILCA to assess the potential of agroforestry in West African farming systems and will assist a proposed Alley Farming Network.

The Alley Farming Network, to be coordinated by IITA with the help of ILCA and ICRAF, will screen multipurpose trees, conduct research on management techniques, and evaluate the fodder potential of different species. This network is an outgrowth of ongoing research on alley cropping at IITA's facilities near Ibadan. IITA and ILCA scientists have been researching various tree and food crop combinations, aligned in rows, since the early 1980s. One promising tree and annual crop partnership tested by IITA and now being adopted by farmers involves one row of leucaena interspersed with several rows of maize. Leucaena regrows easily after cutting, sprouts luxuriant, nutritious fodder, and enriches the soil with nitrogen.

The number of material exchange networks is growing, particularly those involved in screening crop germplasm. International nurseries for the major seed crops are already in place, but many minor crops, root crops, and some industrial crops have thus far been bypassed by networks. In part this is because of the understandable need to concentrate on the major food crops first, but it is also a reflection of the historical difficulty in ensuring that vegetative planting materials destined for export are free of diseases and pests. New technologies for cleaning up live plant materials are now making it much safer to exchange root crop germplasm in tissue culture form. Accordingly, a Southeast Asian Cassava Network is in the planning stage. With assistance from CIAT in Colombia, this network will span Indonesia, Malaysia, the Philippines, and Thailand. IITA has developed a proposal for a Central and West African Root Crops Collaborative Research Network (CEWARCCRN). Vigorous development of networks in tropical perennial crops, particularly fruit, beverage, oil, and fiber crops, is expected in the near future to raise and sustain yields of such commercially important crops as cacao, coffee, oil palm, and mango.

As an outgrowth of an international meeting in Bogor, Indonesia, a decision was reached to start a network focusing on the needs and opportunities for ecophysiological research in Southeast Asia (Osmond, 1986). This network, which will probably begin as an independently planned venture, will harness the skills of ecologists, agronomists, foresters, and scientists in other disciplines. Information generated by this network is expected to help in crop production as well as managed forests that are better adapted to the myriad environments in the region.

Information exchange networks are also growing. For example, IRRI is proposing to help establish a global network for multilanguage communication of agricultural research (Cabanilla and Hargrove, 1986). The idea for this network grew out of a workshop held at IRRI's headquarters near Los Baños, Philippines, in 1983, which was attended by sixty-three participants from Asia, Africa, and Latin America (IRRI, 1984).

The Multilanguage Communication Network for Agriculture plans to issue a newsletter to inform interested individuals about relevant printed materials and reports for adaption, translation, and publication. For each potential item for translation, the newsletter intends to provide information on the objectives and audience as well as the countries and agroclimatic zone or areas for which the methodology or results apply. The network hopes to maintain a computerized data bank of translators and agricultural agencies interested in multilanguage publishing. Materials would be available for agricultural improvement agencies and the private sector in developing countries.

IRRI already has considerable experience in multilanguage publishing and will serve as a model for the network. As of February 1986, thirty-one IRRI books had been published in thirty-four languages in twenty-two countries. IRRI provides text and plates of its publications to interested parties but does not subsidize foreign language publications of its products nor does it charge royalties. Unlike most information exchange networks, the proposed multilanguage network is expected to hold regular workshops for participants in Africa, Latin America, and Asia.

Another information exchange network in the early planning stage will focus on fungal cultures. With a $120,000 grant from the European Economic Community, Commonwealth Agricultural Bureaux International will establish a node in the United Kingdom for a network of European data banks to be known as MINE (Microbial Information Network for Europe). CABI's mainframe computer has information on some eleven thousand strains of fungi maintained in the United Kingdom. MINE will link data bases in the United Kingdom, Netherlands, Federal Republic of Germany, and Belgium (CABI, 1986). Portugal and France are also expected to establish nodes in the network, which will serve agriculture, industry, and medicine.

Several proposed networks deal with crop fertilization and soil fertility. IRRI is pondering whether to set up a blue-green algae network (Roger et al., 1985). Blue-green algae can be free-living in water or live in symbiosis with other organisms. For example, blue-green algae live in leaves of *Azolla* water ferns, which often float in rice paddies. When the water ferns decompose, nitrogenous compounds are liberated and become available to rice plants. IRRI sent a questionnaire to 384 scientists working on various aspects of blue-green algae in fifty-nine countries. Three-quarters of those responding indicated an interest in joining a research network focusing on the microorganisms. Scientists answering the questionnaire expressed a strong desire to standardize field research methods for blue-green algae, to undertake more ecological surveys, and to conduct field experiments under a range of agroecological conditions. A collaborative research network is perceived as the optimal framework for reaching such goals. If this network materializes, strong links are envisaged with INSURF.

Another network dealing with microbiological processes in soil fertility was proposed in 1984 by the International Union of Biological Sciences (Brown and Wolf, 1985). This large-scale collaborative research effort hopes to determine management options for improving soil fertility in the tropics. A similar goal lies behind the International Benchmark Sites Network for Agrotechnology Transfer (Appendix 1), one of the collaborative research networks discussed in Chapter 9.

Although the proposed microbiological processes network on soil fertility has a narrower focus than IBSNAT, the distinction between IBSNAT and another soil fertility network, IBSRAM, is less clear. The Australian Development Assistance Bureau played a leading role in the initiative to launch IBSRAM, which is similar in rationale to IBSNAT but appears to have a more ambitious agenda. IBSRAM is envisaged as a global clearinghouse for information on soils work and is expected to assist national programs in setting up their soils research programs and organizing workshops and symposia. In addition to ADAB, a number of donors have expressed interest in supporting IBSRAM, including the German Agency for Technical Cooperation (GTZ), the Canadian International Development Agency, Britain's Overseas Development Administration, the French Agency for Overseas Research (ORSTOM), and the World Bank.

Network Overload

The surge, perhaps plethora, of networks on the drawing board poses some challenging policy questions. Have we reached network overload in the agricultural sciences? Can communication be improved between networks? Are some networks hanging on too long when they should have closed some time ago? Are networks an adequate substitute for new international centers?

Networking, at least on an international scale, is relatively new in the agricultural sciences so it is difficult to assess whether too many networks are in operation. But in some areas, a surfeit of networks may be leading to crossed wires and duplication of effort. In Africa, where so many networks have been launched, a real danger exists of overloading already relatively weak national programs (Oram, 1988). At least three international networks are devoted to studies of the classification, fertility, and management of tropical soils, including crop modeling. The Soil Management CRSP (Appendix 1), also known as Tropsoils, was started in 1981 and concentrates on soil management to alleviate erosion and degradation in Brazil, Cameroon, Indonesia, Mali, Niger, and Peru. Tropsoils is pursuing such topics as alternatives to shifting cultivation by adopting alternative cropping systems, improved land-clearing practices, and efficient fertilizer schemes for infertile soils. IBSNAT, discussed in more detail in Chapter 9, is expanding its fertilization and crop modeling studies and will probably do so for years to come. Meanwhile, a crop modeling network linking six Asian nations was started in 1986 (D. J. Greenland, pers. comm.). Participants in this nascent network receive instruction at Wageningen, Netherlands, and take IBM PCs back to their institutions. It remains to be seen whether this new network duplicates some of IBSNAT's work.

IBSRAM has not worked closely with IBSNAT, an older, kindred network. In the past this may have been because of the new network's desire to establish its own identity. But lack of collaboration between IBSRAM and IBSNAT may have a price later. IBSRAM, for example, is just beginning to get its networks under way and is testing the network approach. Also, because soils research depends on a common understanding of the soils studied, a possible lack of a common classification system could complicate comparison of research results generated by the analogous networks.

In cases of possible overlap, the argument can be made that each network has a separate mission. Still, some of the new networks impinge strongly on the operations of existing collaborative efforts. Better communication between prospective and existing networks, and between networks already under way, would certainly further research efforts. The Asian Rice Farming Systems Network has taken on long-term fertilizer trials, but no collaboration has developed with INSURF on this component of the network, even though both networks are coordinated by IRRI and are more than a decade old.

Donors are partly to blame for network overlaps. As outlined in Chapter 11, little coordination has emerged between supporters of international agricultural research, and the special agendas of funding agencies can lead to duplication in the field. An exception is the work of the Special Program for African Agricultural Research, which is attempting to coordinate support and rationalize networks in Africa. Research directors and scientists need to be more vigilant to reduce redundancy. International centers could also work more closely together to further network linkages (Walsh, 1986b). The concerted effort of four international centers in the AFRENA agroforestry networks in Africa is a good example of what can be done to harmonize the efforts of international centers.

Several networks work closely together and can serve as models of cooperation. PRECODEPA, for example, is sharing information and technology regarding tuber moth and late blight disease with a sister potato research network, PRACIPA (Appendix 1), which covers the Andes. PRECODEPA training courses also service some sister regional potato networks. For example, the two-month PRECODEPA course on potato production held in Toluca, Mexico, in 1988 had six participants from PRACIPA, the Andean regional potato network (three trainees from Bolivia and one each from Colombia, Peru, and Venezuela). The two-week PRECODEPA course on germplasm management, also held in Mexico in 1988, trained six scientists from PRECODEPA countries and two each from PRACIPA and PROCIPA (a southern cone potato network). The proposed blue-green algae network is exemplary in its determination to complement INSURF's more ambitious program. INSURF, in turn, operates an acid soil lowland rice nursery in conjunction with INGER. This nursery has been set up to identify rice lines that tolerate strongly acidic soils with or without various amendments (Mamaril, 1984). This specialized nursery began in 1981 with ten entries and grew to ninety-nine submissions by 1984. MIRCENs and the multinational Nitrogen Fixation by Tropical Agricultural Legumes (NifTAL) run by the University of Hawaii have also forged helpful linkages in research on biological nitrogen fixation.

Most networks in international agricultural research are less than a decade old so it may be premature to pass judgment on whether some of them have outlived their usefulness. From ten to twenty years normally elapse between the time research begins on a project and when a product is ready for farmers (Plucknett and Smith, 1986b). One of the central ideas of networks is that they are flexible and can be easily disbanded when their task is complete. Nevertheless, the temptation to hang on, perhaps to conduct follow-through projects, can be strong. Evolution does not always lead to perfection. We must hope that donors and administrators will exercise discipline here. Some networks, such as many international nurseries, can be justified on a long-term basis, but others, such as those set up to solve narrowly defined problems, should close down at the completion of their missions or if no progress is made.

To a certain extent, networks are substituting for new international agricultural research centers. INIBAP is a good example of a network that was set up because no international center for research on bananas and plantains was operating or was likely to. International agricultural research centers each may require between $5 and $25 million annually for their operations, depending on the research agenda, and start-up costs are considerable. Virtually all of the international agricultural research centers were created in the 1960s and 1970s to pursue research on major food crops and livestock in the Third World, but that bow wave of institution building has passed

(Plucknett and Smith, 1982). Some fifteen to twenty sizable international centers are currently pursuing agricultural research, but few new ones are likely soon, given the cost of establishing and maintaining such centers. Skepticism has been voiced that some of the existing international centers are too big, and more of these institutions may not necessarily lead to higher agricultural yields.

A pattern is emerging in which networks are seen as cheap and relatively quick solutions to research problems. New international centers may be the ideal in some cases, but most will never come to pass given the current tight fiscal climate for research. Provided that sufficient attention is given to ensuring that institutions within a network are fortified so that they can participate adequately, networks are likely to continue to be a cost-effective format for research. In many cases, networks probably can substitute for new international centers if network nodes are upgraded with improved facilities and better-trained personnel and by ensuring that coordination is in competent hands. Some networks as international centers (NIARCs) may well be formed to meet such research needs.

Strengthening of National Programs

As national programs in the Third World strengthen, they will increasingly take over much of the planning and coordination of networks. International centers are likely to remain heavily involved in setting up and managing networks in some regions, however, particularly Sub-Saharan Africa (Brady, 1985). Progress in overcoming Africa's food woes will require sizable investments in training personnel and upgrading research facilities (Asante, 1986). Two dimensions of networking in which international centers have a comparative advantage over national programs are facilities and personnel suitable for research training and information management. These advantages are likely to hold for some time to come.

International centers will continue to host or organize many of the courses and intensive workshops tailored to specific networks. Other courses offered by international centers will also benefit current and future networks by building skills in a broad range of disciplines.

International centers, particularly those under the CGIAR umbrella, are strongly committed to training Third World nationals, and many plan to increase their support for such activities.

By 1984, 19,450 individuals from 117 countries had received training at the thirteen CGIAR centers, or were supported for university studies by the international centers, up from 16,427 in 1983 (Bunting, 1985; TAC, 1986b). Almost one-fifth of those trainees participated in courses lasting at least nine months. The majority of trainees participated in short courses lasting a month or two. Of the 19,450 trainees, 640 were postdoctoral colleagues and 3,000 were university students. CGIAR centers have spent $90 million on training, exclusive of indirect costs and adjustments for inflation, during the last two decades. On average, CGIAR centers devote 8 percent of their budgets to training and this proportion is growing. ILCA increased its training budget from $0.5 million in 1981 to $1.1 million in 1986 (Brumby, 1986). In 1986, training accounted for 5 percent of ILCA's budget, but in 1989 it accounted for 9 percent of the center's growing budget (ILCA, 1988).

IRRI has emphasized training since its inception and is the leading international center in the training of scientists and technicians (Table 19). The number of IRRI alumni, now dispersed in more than seventy countries with many participating in networks (Figure 17), continues to grow. As of 1983, for example, IRRI had trained 3,451 people; by November 1984 that figure had reached 3,700 and by May 1985, the ranks of IRRI trainees had swollen to 4,006 (Bunting, 1985). About 80 percent of IRRI trainees are from Asia, where rice is a basic staple.

Table 19. Numbers of people trained at, or sponsored by, international centers within the Consultative Group on International Agricultural Research (CGIAR)

Center	Period	Persons trained
IITA	1972–86	6,000
IRRI	1962–83	3,451
CIMMYT	1966–86	3,000
CIAT	1969–83	2,459
WARDA	1973–83	1,128
IBPGR	1979–83	898
ICRISAT	1974–83	825
CIP	1978–83	587
ILRAD	1978–83	432
ICARDA	1975–83	362
ILCA	1975–83	340

Sources: Bunting, 1985; Hepworth, 1987; Oyekan, 1987.

Note: See Appendix 3 for acronyms and research missions of the centers.

Figure 17. An IRRI-trained cropping systems specialist employed by the Philippines national program and participating in on-farm, cowpea-rice rotation trials of the Asian Rice Farming Systems Research Network, Pangasinan Province, Philippines, June 1986.

IRRI offers courses on statistics, rice genetic resources, rice genetic evaluation and utilization, cropping systems, upland rice, agricultural communication, integrated pest management, irrigation water management, and agricultural economics (Pathak, 1985).

CIMMYT is another major training center for Third World agricultural scientists. It has trained nearly nine hundred maize specialists from seventy-four countries since 1971 (Breth, 1986). CIMMYT trainees are not kept all day in air-conditioned lecture halls and laboratories. Long stretches in the field are an early baptism for aspiring maize and wheat breeders. Trainees are taught to recognize desirable traits in crop lines and to conduct scientifically sound trials in good weather or bad. On-farm testing is one of the cornerstones of CIMMYT's philosophy so trainees gain experience under real farming conditions as well as at experiment station sites. CIMMYT inaugurated a new training wing at its headquarters in Mexico in 1987.

The impressive number of scientists from developing countries going to international centers for training constantly strengthens network bonds and sets the stage for future collaborative efforts. ICRISAT, with its global mandate for research on sorghum, pearl millet, pigeonpea, chickpea, and groundnut, has a relatively modest training program (Table 8), yet the India-based center supported nine hundred trainees from sixty-nine countries between 1974 and 1984 (ICRISAT, 1985). And the International Fertilizer Development Center, which works closely with soil networks such as INSURF, has trained between 104 and 426 people annually between 1979 and 1984. The Alabama-based center's course offerings are divided into two main groups: fertilizer marketing, and fertilizer production and technology (IFDC, 1985).

The Collaborative Research Support Programs that serve mutual interests of some U.S. universities and national programs in developing countries have also made an important contribution to the upgrading of agricultural scientists in the Third World. With support from USAID, the various CRSPs (Appendix 1) have arranged graduate training for more than five hundred students from developing countries at U.S. universities (Hogan et al., 1986).

Computer Networking

Another dimension of networking in which international centers maintain a comparative advantage is in the area of communications and data-base management. Some writers have characterized the late 1970s and early 1980s as the beginning of the information age, rivaling the importance of the earlier industrial and scientific revolutions. Networks of computers are increasingly being used to nurture networks of people (Denning, 1987). With the quickening pace of technological, ecological, and socioeconomic change, decision makers increasingly need reliable, up-to-the-minute information upon which to base their decisions. The efficiency of information collection, synthesis, and retrieval is thus of paramount importance to networks, and the international agricultural research centers will always play a major role in these tasks.

New machines cannot substitute for innovative thinking, but they provide more time for creative tasks. Just as international and some national agricultural research centers employ computers to streamline administration and germplasm conservation work, so will scientists increasingly adopt breakthroughs in microelectronics to improve the efficiency of research and communication.

In discussing the role of computers in networking, P. J. Denning (1987) identifies five evolutionary stages in the process of tighter integration and fuller collaboration between participants: file transfer, remote connections, distributed computation, real-time collaboration, and coherent function. At the first stage, files are transferred between participants and computers are used for electronic mail and bulletin boards. In the second stage, users can tap into remote sources such as data bases. During the third stage, the computer network is able to support computing processes at widely scattered nodes. At the fourth stage, the network permits real-time conferences for users at different workstations in which they can jointly share information, edit text, run programs, and examine results. Finally, in the fifth stage, the network is a coherent system in which people contribute and use their resources.

A diverse range of computers and satellite technologies for speeding up data processing and information is on the market, but the infrastructure necessary for their implementation is still inadequate in many Third World areas. Computers can exchange information over telephone lines, but such links are sporadic and often function erratically in many developing countries. Many of these countries are improving their telecommunications systems, however, and such enhancements will facilitate research networks.

Local, regional, and national computer networks are already common in most developed countries. In North America, for example, BITNET links computers at sixty U.S. and eight Canadian universities. By 1987, the European Academic and Research Network (EARN) had connected some six hundred computers in twenty countries, including Iceland, Israel, and Côte d'Ivoire, an impressive accomplishment in three years of operation (Dickman, 1987; Jennings, 1987). Approximately thirty thousand academics use EARN's three basic services—file transfer, electronic mail, and remote job execution—to further their research. Both networks have formed since 1981 and can now communicate with each other via Darmstadt, Federal Republic of Germany, or Montpellier, France.

BIONET is a specialized computer network for biologists and biochemists in North America and Europe supported by the National Institutes of Health and subscription fees (Kristofferson, 1987). The 1,750 subscribers to BIONET have access to data bases, powerful software for analysis of DNA and proteins, a bulletin board, and electronic mail. Scientists in industrial nations use a variety of computer networks to tap data bases and exchange information. These networks can serve as models to further agricultural research networks in the Third World.

By 1987, all thirteen international centers within the CGIAR organization were linked by electronic mail, a communication system that is as quick as telex but much less expensive when communication traffic is heavy. A large central computer contains numerous "mailboxes" for individuals or institutions to which messages can be sent or retrieved in seconds. CGIAR centers use electronic mail to exchange messages between each other and with a large variety of organizations involved in conducting or supporting agricultural research, such as IDRC, Winrock International, and several U.S. universities.

Other agricultural institutions involved in networks are likely soon to install computer terminals at various locations. Many of these terminals will eventually be able to communicate with computers at other institutions, both nationally and abroad. Some problems remain to be ironed out before computer networking is truly operational on a global scale. Research institutions currently contain a bewildering array of computer makes, many of which are incompatible. Even models made by the same computer manufacturer are not always able to work together smoothly. Software is even more diverse than computers and printers. Scientists are often burdened with complex procedures required to access electronic mail or data bases (Jennings et al., 1986). Just one section of the World Bank had to struggle for weeks trying to install a computer network (Coronel, 1987).

New technologies understandably present fresh challenges, but they can be overcome. With perseverance and ever faster and more powerful computers, computer networking will increase the efficiency of research. Computer manufacturers have become sensitive to clients' demands that the various mix of mainframes and microcomputers work together more effectively. The number of computer manufacturers is declining, and there is a tendency to adopt a common model so the issue of incompatability is likely to decrease in the future.

As computer facilities become better integrated on a global scale, more network collaborators will be able to call up data obtained elsewhere in a fraction of the time it takes to use the mail. Teleconferences, using computers and satellite dishes, will permit network participants to conduct some workshops without the expense, inconvenience, and time loss associated with travel. Computer prices are plummeting so Third World institutions are increasingly able to acquire terminals, multicolored plotters, and rapid printers. India, for example, is establishing a countrywide computer data base with the aid of a satellite. This $70 million system will provide access to data bases for government workers in 430 districts, the 22 state capitals, and other major cities (Jayaraman, 1987). These data bases, containing information on health, agriculture, rural development, employment, energy consumption, and other topics, will be used particularly by development planners. In view of the growing power and potential of computer networks, international centers are rapidly adding computer data-base management and other computer applications to their training programs.

International centers will continue to play a catalytic and service role for networks for decades to come. The leadership and coordination of networks, however, will be increasingly taken over by stronger national programs on a rotational basis. Those networks in which international centers retain a central position as coordinators and planners would benefit from even greater sensitivity to the needs of national programs. M. S. Swaminathan, former director general of IRRI, points to the need for national agricultural programs to act more like partners in the planning of international networks (CGIAR, 1987).

In the near future, the absolute number of networks is likely to continue to rise steeply. Some networks will undoubtedly break up either because they have accomplished their missions or because donors withdrew support, but the rush toward networking shows no sign of abating. Eventually, the number of networks in agricultural research will plateau and perhaps ultimately dip, but not in the foreseeable future.

Some critics of plant breeders' rights and the commercialization of agriculture have suggested that research collaboration will wane as the profit motive increasingly takes over. But even in industrial countries, public institutions still produce the bulk of varieties of many important crops, and private seed companies have barely established a foothold in most Third World nations. Even though hybrid seed production for a few crops is increasing worldwide, many government-run operations produce hybrid maize and other crops for farmers. Plant breeders' rights only protect the seed companies from others selling the proprietary seed; protected varieties can be used in further crossing and entered in nursery trials. The immediate parental lines of hybrids are closely guarded, but the genes that went into producing those parental lines are available in other germplasm. International nurseries will remain important testing grounds for protovarieties. More international nurseries will be set up to tackle problem environments and to cater to the needs of breeders working with specialty crops and plants of regional importance such as some tropical fruit trees and unique Andean food grains.

As the number of networks continues to climb, accountability will become an increasingly pressing issue. We have identified some of the principles for success, pitfalls to be avoided, management features that have enhanced some networks, and specific products arising out of networks, but more research is warranted in analyzing the performance of networks. The field is ripe for agricultural economists and management specialists to conduct in-depth surveys of selected networks and to devise appropriate methodologies for assessing the adequacy of network operations more precisely. One area requiring particular attention is the financial savings of conducting collaborative research compared to independent, isolated projects. Collaborative research is widely regarded as highly beneficial, but the argument for cooperation would be more convincing if hard data were available to demonstrate monetary benefits. Network products need to be assessed in financial and social terms so that donors will be better informed when pondering grants for networks.

Appendix 1

Some International Agricultural Research Networks, the Year Started, the Coordinating Institution, the Nations Participating, and the Regional Coverage

Network	Year begun	Coordinator*	Nations	Countries/region
AFRENA (Agroforestry Research Networks for Africa)[1]	1985	ICRAF	13	Africa
AGLN (Asian Grain Legume Network)[1]	1986	ICRISAT	10	Asia
Animal Traction Research Network[2]	1988	ILCA	25	Africa
ARFSN (Asian Rice Farming Systems Network)[1]	1975	IRRI	15	Asia, Madagascar
ARNAB (African Research Network for Agricultural Byproducts)[3]	1981	ILCA	3	Cameroon, Nigeria, Senegal
ARSHAT (Asian Regional Sorghum Hybrid Adaptation Trial)	1986	ICRISAT	3	India, Pakistan, Thailand
Cattle Milk and Meat Network[1]	1988	ILCA	8	West and Central Africa
CIMMYT Central America and Caribbean Regional Maize Program[3]	1973	CIMMYT	13	Central America, Caribbean
CIMMYT East Africa Regional Maize Program[3]	1980	CIMMYT	8	East Africa
CIMMYT South American Regional Maize Program[3]	1976	CIMMYT	7	South America
CIMMYT South and South East Asia Regional Maize Program[3]	1976	CIMMYT	15	South and Southeast Asia

*See Appendix 2 for acronyms of coordinating institutions.
[1]Collaborative research network.
[2]Information exchange network.
[3]Scientific consultation network.
[4]Material exchange network.
#Coordinator varies according to subnetwork or location of annual meeting.

Network	Year begun	Coordinator	Nations	Countries/region
CIMMYT Southern Africa Regional Maize Program[3]	1985	CIMMYT	7	Southern Africa
CIMMYT/ESA (CIMMYT Eastern and Southern Africa Economics Program)[3]	1976	CIMMYT	15	Africa
Confectionary Groundnut Varietal Trial	1985	ICRISAT	14	World
CRSP—Bean/Cowpea (Collaborative Research Support Program—Bean/Cowpea)[1]	1980	MSU	14	Africa, United States, Latin America
CRSP—Fisheries Stock Assessment[1]	1985	UM	3	Costa Rica, Philippines, United States
CRSP—Peanut[1]	1982	UG	14	Africa, United States, Southeast Asia, Caribbean, Central America
CRSP—Pond Dynamics[1]	1982	OSU	6	Central America, Africa, United States, Southeast Asia
CRSP—Small Ruminants[1]	1978	UCD	6	South America, Africa, United States, Southeast Asia
CRSP—Soil Management[1]	1981	NCSU	7	Africa, United States, South America, Southeast Asia
CRSP—Sorghum/Millet[1]	1979	UN	10	Southeast Asia, United States, Africa, Latin America
EACSSN (Eastern Africa Cooperative Sorghum Screening Nursery)[4]	1985	ICRISAT	8	East Africa
EARSAM (Eastern Africa Regional Sorghum and Millets Network)[1]	1982	ICRISAT	8	East Africa
ESARRN (East and Southern Africa Root Crop Network)[1]	1985	IITA	8	Africa
FADINAP (Fertilizer Advisory Development and Information Network for Asia and the Pacific)	1978	ESCAP	24	Asia, Pacific
Groundnut Early Maturing Varietal Trial	1985	ICRISAT	5	Asia, Africa
IBSNAT (International Benchmark Sites Network for Agrotechnology Transfer)[1]	1982	UH	26	World
IBSRAM (International	1983	IBSRAM	20	World

Network	Year begun	Coordinator	Nations	Countries/region
Board for Soils Research and Management)[1]				
IBYAN (International Bean Yield Adaptation Nursery)[4]	1976	CIAT	30	World
ICAT (International Chick-pea Adaptation Trial)[1]	1981	ICRISAT	3	Colombia, Korea, Cape Verde
INGER (International Network on Genetic Enhancement of Rice)[1,4]	1975	IRRI	80	World
INIBAP (International Network for the Improvement of Banana and Plantain)[1]	1984	INIBAP	13	Latin America, Caribbean, Southeast Asia
INSURF (International Network on Soil Fertility and Sustainable Rice Farming)[1]	1976	IRRI/IFDC	21	Asia, Africa, Caribbean, South America
International Maize Improvement Network[4]	1970	CIMMYT	90	World
International Wheat Nursery System[4]	1960	CIMMYT	115	World
INTSOY (International Soybean Program)[1,4]	1973	UIUC	132	World
IPMAT (International Pearl Millet Adaptation Trial)[4]	1975	ICRISAT	8	South Asia, Africa
IPMDRTP (International Pearl Millet Disease Resistance Testing Program)[4]	1976	ICRISAT	4	South Asia, Africa
ISVAT (International Sorghum Variety Adaptation Trial)[4]	1977	ICRISAT	37	World
IWBP (International Witches' Broom Project)[1]	1985	UF	5	South America, Trinidad
MIRCENs (Micro-Biological Resources Centers)[3,4]	1976	UNEP/ UNESCO	20	Africa, Asia, United States, Latin America
NifTAL (Nitrogen Fixation by Tropical Agricultural Legumes)[1,4]	1975	UH	53	World
ONESASA (Oilcrops Network for Eastern and Southern Africa and South Asia)[3]	1981	IAR	11	Africa, South Asia
PANESA (Pasture Network for Eastern and Southern Africa)[1]	1984	ILCA	19	Africa
PESTNET (African Regional Pest Management Research and Development Network for Integrated Control of Crop and Livestock Pests)[3]	1986	ICIPE	10	Africa
PIN (Pigeonpea International Nurseries)[4]	1975	ICRISAT	22	World

Network	Year begun	Coordinator	Nations	Countries/region
PRACIPA (Programa Regional Andino Cooperativo de Papa)[1]	1982	CIP	5	Andean countries
PRAPAC (Programme Regional d'Amelioration de la Culture de Pomme de Terre en Afrique Centrale)[1]	1982	CIP	4	Central Africa
PRECODEPA (Programa Regional Cooperativo de Papa)[1]	1978	CIP	10	Central America, Caribbean
PROCIPA (Programa Cooperativo de Investigaciónes en Papa)[1]	1982	CIP	4	Southern South America
Public Awareness Association for International Agricultural Research[2]	1988	CGIAR	10	World
RISPAL (Rede de Investigación en Sistemas de Producción Animal en Latinoamerica)[3]	1986	IICA	13	Americas
RNAM (Regional Network for Agricultural Machinery)[4]	1977	ESCAP	10	Asia
RRS (Recherche sur Résistance a Sécheresse)[3]	1984	CIRAD/ ORSTOM	9	Africa, Canada, France
Sago Advancement Group Office[2]	1985	ESPG	13	Europe, Southeast Asia, Japan, Australia
SAPPRAD (Southeast Asian Program for Potato Research and Development)[1]	1980	CIP	5	Southeast Asia, Sri Lanka
SDI—ICRISAT (Selective Dissemination of Information—ICRISAT)[2]	1988	ICRISAT	37	Semiarid tropics
SDI—ILCA[2]	1983	ILCA	32	Africa
Small Ruminant and Camel Group Research Network[3]	1986	ILCA	9	Africa
Sorghum Breeding Network[4]	1982	ICRISAT	40	Semiarid tropics
SUAN (Southeast Asian Universities Agroecosystem Network)[3]	1982	#	4	Southeast Asia
Trypanotolerant Livestock Network[1]	1981	ILCA	10	Africa
WAFSRN (West African Farming Systems Research Network)[2,3]	1982	ABU	17	West Africa
West Africa Groundnut Collaborative Network[3]	1987	ICRISAT	11	West Africa
West and Central African Cowpea Improvement Network[3]	1987	IITA	16	West and Central Africa

Appendix 2

Acronyms and Locations of Institutions or Organizations Serving as Coordinators of International Agricultural Research Networks

Acronym	Name	Location
ABU	Amadu Bello University	Zaria, Nigeria
CGIAR	Consultative Group on International Agricultural Research	Washington, D.C.
CIAT	Centro Internacional de Agricultura Tropical	Cali, Colombia
CIMMYT	Centro Internacional de Mejoramiento de Maiz y Trigo	El Batán, Mexico
CIP	Centro Internacional de la Papa	Lima, Peru
CIRAD	Centre International de Recherche Agricole et le Développement	Montpellier, France
ESCAP	Economic and Social Commission for Asia and the Pacific	Los Baños, Philippines/ Bangkok, Thailand
ESPG	East Sepik Provincial Government	Wewak, Papua New Guinea
IAR	Institute of Agricultural Research	Addis Ababa, Ethiopia
ICIPE	International Center of Insect Physiology and Ecology	Nairobi, Kenya
ICRAF	International Council for Research in Agroforestry	Nairobi, Kenya
ICRISAT	International Crops Research Institute for the Semi-Arid Tropics	Patancheru, A.P., India
IFDC	International Fertilizer Development Center	Muscle Shoals, Alabama
IICA	Inter-American Institute for Cooperation in Agriculture	San José, Costa Rica
IITA	International Institute of Tropical Agriculture	Ibadan, Nigeria
ILCA	International Livestock Center for Africa	Addis Ababa, Ethiopia
IRRI	International Rice Research Institute	Los Baños, Philippines
MSU	Michigan State University	East Lansing, Michigan
NCSU	North Carolina State University	Raleigh, North Carolina

Acronym	Name	Location
ORSTOM	Office de Recherche Scientifique Outre-Mer	Paris, France
OSU	Oregon State University	Corvallis, Oregon
UCD	University of California, Davis	Davis, California
UF	University of Florida	Gainesville, Florida
UG	University of Georgia	Experiment, Georgia
UH	University of Hawaii	Honolulu, Hawaii
UIUC	University of Illinois at Urbana-Champaign	Urbana, Illinois
UM	University of Maryland	College Park, Maryland
UN	University of Nebraska	Lincoln, Nebraska
UNEP	United Nations Environment Program	Nairobi, Kenya
UNESCO	United Nations Educational, Scientific, and Cultural Organization	Paris, France

Appendix 3

International Agricultural Research Centers Involved in Agricultural Research Networks*

Acronym	Institution	Year founded	Research programs
AVRDC	Asian Vegetable Research and Development Center	1971	Soybean, sweet potato, tomato, mungbean, Chinese cabbage, hot and sweet pepper
CIAT	Centro Internacional de Agricultura Tropical	1968	Cassava, beans, rice, pastures
CIMMYT	Centro Internacional de Mejoramiento de Maiz y Trigo	1964	Maize, wheat, triticale, barley
CIP	Centro Internacional de la Papa	1972	Potato, sweet potato
IBPGR	International Board for Plant Genetic Resources	1973	Plant genetic resources
IBSRAM	International Board for Soils Research and Management	1983	Soil management practices for developing countries
ICARDA	International Center for Agricultural Research in the Dry Areas	1976	Wheat, barley, triticale, faba bean, lentil, chickpea, forages
ICIPE	International Centre of Insect Physiology and Ecology	1970	Livestock ticks, tsetse flies, crop pests
ICLARM	International Center for Living Aquatic Resources Management	1977	Fisheries, aquaculture, coastal area management
ICRAF	International Council for Research in Agroforestry	1978	Agroforestry

*See Map 1 for locations of institutions.

Acronym	Institution	Year founded	Research programs
ICRISAT	International Crops Research Institute for the Semi-Arid Tropics	1972	Chickpea, pigeonpea, pearl millet, sorghum, groundnut
IFDC	International Fertilizer Development Center	1974	Fertilizer formulations and optimal application methods
IFPRI	International Food Policy Research Institute	1975	Food and agricultural policy
IIMI	International Irrigation Management Institute	1984	Irrigation system management
IITA	International Institute of Tropical Agriculture	1967	Maize, rice, cowpea, sweet potato, yams, cassava, soybean
ILCA	International Livestock Center for Africa	1974	Livestock production and husbandry
ILRAD	International Laboratory for Research on Animal Diseases	1974	Trypanosomiasis, East Coast fever
INIBAP	International Network for the Improvement of Banana and Plantain	1980	Banana, plantain
IRRI	International Rice Research Institute	1960	Rice breeding, pests, agronomy
ISNAR	International Service for National Agricultural Research	1980	Strengthening national agricultural research systems
ITC	International Trypanotolerance Centre	1984	Trypanotolerance in cattle, especially N'Dama breed
WARDA	West Africa Rice Development Association	1971	Rice

Note: Many of the international agricultural research centers also have economics or farming systems research programs.

Appendix 4

International Network on Genetic Enhancement of Rice (INGER) Entries Named as Varieties in Different Countries as of 1985

Country	Designation	Origin	Named	Year released
		East Asia		
China	Chianung sen yu 23	Taiwan	—	1981
	Chianung si-pi-661020	Taiwan	—	1981
	Suweon 290	Korea	—	1981
	IR1541-102-7	IRRI	IR26	1980
	IR2061-214-3-8-2	IRRI	IR28	1980
	IR1561-228-3-3	IRRI	32 Xuan 5	1980
	BG90-2	Sri Lanka	BG90-2	1983
		Southeast Asia		
Burma	BG90-2	Sri Lanka	Sinthiri	1978
	BKN6986-108-3	Thai/IRRI	Yenet-1	1980
	BKN6986-167	Thai/IRRI	Yenet-2	1981
	BR51-91-6	Bangladesh	Sintheingi	1978
	C4-113	Philippines	Sein TaLay	1979
	C22	Philippines	Yar-1	1981
	IR28	IRRI	Shwe-War-Lay	1977
	IR34	IRRI	Sinshwethwe	1978
	KN96	Indonesia	Yar-2	1981
	KN117	Indonesia	Yar-3	1981
	Kulu	Australia	Sin Kalaya	1980
	LG240	Indonesia	Yar-4	1981
	IR1529-680-3-2	IRRI	Yar-5	1981
	Pelita I-1	Indonesia	Palethwe	1979
	IR751-592	IRRI	Shwethwelay	1979
	IR42	IRRI	Pyi Lone Chan Tha	1983
	IR21836-90-3-3	IRRI	Hmawbi-2	1983
	BH2	Sierra Leone	Yenet-5	1983
	RD19	Thailand	Yenet-6	1983
	B922c-Mr-118	Indonesia	Yenet-7	1983

Country	Designation	Origin	Named	Year released
Indonesia	IR26	IRRI	IR26	—
	IR28	IRRI	IR28	—
	IR30	IRRI	IR30	—
	IR32	IRRI	IR32	—
	IR34	IRRI	IR34	—
	IR36	IRRI	IR36	1978
	IR38	IRRI	IR38	1978
	IR42	IRRI	IR42	—
	IR50	IRRI	IR50	1981
	IR52	IRRI	IR52	1981
	IR54	IRRI	IR54	1981
	IR2071-621-2	IRRI	Asahan	1978
	IR2307-247-2-2-3	IRRI	Semeru	1980
	IR46	IRRI	IR46	1983
	IR13543-66	IRRI	Kelara	1984
Kampuchea	IR42	IRRI	—	—
Malaysia	IR42	IRRI	—	—
Philippines	KN-1b-361-1-8-6-10	Indonesia	RP Kn-2	—
Vietnam	IR36	IRRI	NN 3A	1978
	Jaya	India	—	—
	Pelita I-1	Indonesia	—	—
	Biplab	Bangladesh	—	—
	IR2307-247-2-2-3	IRRI	NN 6A	1980
	IR2823-399-5-6	IRRI	NN 2B	1980
	IR1561-228-3-3	IRRI	—	1981
	IR2797-115-3	IRRI	NN 3B	1980
	IR9129-192-2-3-5	IRRI	NN 7A	1980
	IR9224-73-2-2-2-3	IRRI	OM33	1983
	IR8423-132-6-2-2	IRRI	—	1983
	IR42	IRRI	—	—
		South Asia		
Bangladesh	IR2061-214-3-8-2	IRRI	BR6	1978
	IR2053-87-3-1	IRRI	BR7	1978
India	BR51-46-1-Cl	Bangladesh	—	1979
	IR30	IRRI	IR30	1979
	IR34	IRRI	IR34	1979
	IR36	IRRI	IR36	1979
	IR50	IRRI	IR50	—
	Intan	Philippines	Intan	—
	HPU734	IRRI	Himalaya-1	1982
	HPU71	India	Himalaya-2	1982
	IR579-48-1-2	IRRI	IR579	1979
	IR1561-216-6	IRRI	Prasad	1978
	IR13427-45-2	IRRI	Py3	1983
	BG90-2	Sri Lanka	Pant Dhan 4	1983
	CO43	India	—	1983
	CO44	India	—	1983
	IR42	IRRI	AU42/1	1983
	IR1846-284-1	IRRI	VL Dhan 16	1984
Nepal	IET1444	India	Bindeshwari	1981
	BG90-2	Sri Lanka	Janaki	1978
	IET2935	India	Durga	1978
	IR2071-124-6-4	IRRI	Sabitri	1978

Country	Designation	Origin	Named	Year released
	IR3941-4-Plp2B	IRRI/India	Kanchen	1982
	IR2298-PlpB-3-2-1-1B	IRRI/India	Himali	1982
	Mala/J15	Bangladesh	Mallika	1982
	IR2061-628-1-6-4-3	IRRI	Laxmi	1978
Pakistan	IET4094	India	DR-82	1984
	IR2053-261-2-3	IRRI	DR-83	1984
		Southwest Asia		
Iran	IR28	IRRI	Amol 2	1982
	Sona	India	Amol 3	1982
Turkey	Plovdiv	Bulgaria	—	—
	Krosnodoroski 424	USSR	—	—
	Lieto	Italy	—	—
		Africa		
Burkina Faso	IR1529-680-3-2	IRRI	IR1529	1979
	Vijaya	India	—	—
Cameroon	IR42	IRRI	—	—
	IR46	IRRI	IR46	1981
Côte d'Ivoire	Jaya	India	—	—
	IR46	IRRI	—	—
Egypt	IR1561-228-3-3	IRRI	—	1981
Ghana	IR1820-210-2	IRRI	Tamale 1	1978
Kenya	IR1561-228-3-3	IRRI	—	1981
Liberia	IR1416-131-5	IRRI	Suakoko 12	1981
Mali	IR1529-680-3	IRRI	—	—
	Jaya	India	—	—
Mauritania	IR1561-228-3-3	IRRI	IR1561	1981
Niger	IR1529-680-3-2	IRRI	IR1529	1979
Nigeria	IR30	IRRI	IR30	1981
	IR42	IRRI	IR42	1981
	IR46	IRRI	IR46	1981
Senegal	Jaya	India	—	—
Sierra Leone	PRJ12	CIAT	Rok11	—
	PRJ1	CIAT	Rok12	—
Sudan	IR2053-206-1-3-6	IRRI	IR2053	1979
Tanzania	BG90-2	Sri Lanka	—	1984
	IET1444	India	—	1984
	ITA233	Nigeria	—	1984
	ITA212	Nigeria	—	1984
	Pinulot 330	Sri Lanka	—	1984
	IET2397	India	Katrin 1	1984
		Latin America		
Argentina	IR841-63-5-18	IRRI	IR841	1981
Belize	Cica 8	CIAT/ICA	Cica 8	—
Bolivia	IR43	IRRI	Saavedra 5	1981
Brazil	IR442-2-58	IRRI	Arroz BR-2	1978
	IR8208-146-1	IRRI	Pesagro 101	1982
	IR46	IRRI	Pesagro 102	1982
	IR841-67-1-2	IRRI	Empasc 104	1985
	P1278-6-17M-1-1B	Colombia	IAC 1278	1982
	CICA 8	CIAT/ICA	INCA 4440	1982
Colombia	P1035-5-6-1-1-2M	Colombia	Metica 1	1981
	P2023F4-74-2-1B	Colombia	Oryzica 2	1984

Country	Designation	Origin	Named	Year released
Costa Rica	P881-19-22-4-1-1B-CR1	Colombia	CR1821	1985
Cuba	IR43	IRRI	IR1529	1978
Ecuador	IR1545-339-2-2	IRRI	IR1545	1981
Guatemala	Cica 8	CIAT/ICA	ICTA Virginia	—
	P918-25-1-4-2-3-1B	Colombia	ICTA Virginia	1982
Honduras	Cica 8	CIAT/ICA	Cica 8	—
	P918-25-15-2-3-2-1B	Colombia	Yojoa 44	1984
Mexico	SPR6726-134-2-26	Thailand	Cardenas A80	1981
Panama	Cica 8	CIAT/ICA	Cica 8	—
Paraguay	Cica 8	CIAT/ICA	Adelaide 1	—
Peru	IR4570-83-3-3-2	IRRI	PA-3	1984
Venezuela	PR106	India	Araure 3	1984
	P2217F4-30-4-1B	Colombia	Araure 4	1984

Source: Seshu, n.d.

Note: For acronyms and locations of IRRI and CIAT see Appendixes 2 and 3. ICA (Instituto Colombiano Agropecuaria) is Colombia's national agricultural research program. Dashes indicate that the name or year is not known.

Appendix 5. Acronyms

AAASA	Association for the Advancement of Agricultural Sciences in Africa
AAIRNET	African Agricultural Informational Resources Network
ABU	Amadu Bello University
ACIAR	Australian Centre for International Agricultural Research
ADAB	Australian Development Assistance Bureau
ADB	Asian Development Bank
AfDB	African Development Bank
AFRENA	Agroforestry Research Networks for Africa
AFRICALAND	Network on Land Development and Management of Acid Soils in Africa
AGLN	Asian Grain Legume Network
AGRIS	International Information Service on the Agricultural Sciences and Technology
ALAD	Arid Lands Agricultural Development Program
APDC	Asian and Pacific Development Center
ARFSN	Asian Rice Farming Systems Network
ARNAB	African Research Network for Agricultural Byproducts
ARS	Agricultural Research Service
ARSHAT	Asian Regional Sorghum Hybrid Adaptation Trial
ASIALAND	Network on Land Development and Management of Acid Soils in Asia and the Pacific
AVRDC	Asian Vegetable Research and Development Center
BIFAD	Board for International Food and Agricultural Development
BORIF	Bogor Institute for Food Crops
BRITE	Basic Research in Industrial Technologies in Europe
BSP	Benchmark Soils Project
CABI	Commonwealth Agricultural Bureaux International
CAP	Collaborators' Advisory Panel
CATIE	Centro Agronomico Tropical de Investigación y Ensenañza
CCRN	Cooperative Cereals Research Network
CEAREP	CIMMYT Eastern and Southern Africa Economics Program
CERN	European Laboratory for Particle Physics
CEWARCCRN	Central and West African Root Crops Collaborative Research Network
CGIAR	Consultative Group on International Agricultural Research
CIAT	Centro Internacional de Agricultura Tropical

CIDA	Canadian International Development Agency
CIMMYT	Centro Internacional de Mejoramiento de Maiz y Trigo
CIMMYT/ESA	CIMMYT Eastern and Southern Africa Economics Program
CIP	Centro Internacional de la Papa
CIRAD	Centre International de Recherche Agricole et le Développement
COE	Comité Ejecutivo
COPERE	Comité Permanente Regional
CRSP	Collaborative Research Support Program
EACSSN	Eastern Africa Cooperative Sorghum Screening Nursery
EARN	European Academic and Research Network
EARSAM	Eastern Africa Regional Sorghum and Millets Network
EEC	European Economic Community
ELAR	Latin American Rust Nursery
ELVTs	Elite Variety Trials
EMBL	European Molecular Biology Laboratory
EMC	Executive Management Committee
ESARRN	East and Southern Africa Root Crop Network
ESCAP	Economic and Social Commission for Asia and the Pacific
ESO	European Southern Observatory
ESPG	East Sepik Provincial Government
ESYT	Elite Selection Yield Trial
EVTs	Experimental Variety Trials
FADINAP	Fertilizer Advisory Development and Information Network for Asia and the Pacific
FAO	Food and Agriculture Organization
FSAR	Farming Systems Adaptive Research
GTZ	German Agency for Technical Cooperation
IAR	Institute of Agricultural Research
IARCs	International agricultural research centers
IBPGR	International Board for Plant Genetic Resources
IBSNAT	International Benchmark Sites Network for Agrotechnology Transfer
IBSRAM	International Board for Soils Research and Management
IBWSN	International Bread Wheat Screening Nursery
IBYAN	International Bean Yield Adaptation Nursery
ICA	Instituto Colombiano Agropecuaria
ICAR	Indian Council for Agricultural Research
ICARDA	International Center for Agricultural Research in the Dry Areas
ICAT	International Chickpea Adaptation Trial
ICIPE	International Center of Insect Physiology and Ecology
ICLARM	International Center for Living Aquatic Resources Management
ICLCD	International Committee on Land Clearing and Development in the Tropics
ICRAF	International Council for Research in Agroforestry
ICRISAT	International Crops Research Institute for the Semi-Arid Tropics
ICTA	Instituto de Ciencia y Tecnología Agrícola
IDRC	International Development Research Centre
IFAD	International Fund for Agricultural Development
IFDC	International Fertilizer Development Center
IFPRI	International Food Policy Research Institute
IFRON	International Floating Rice Observational Nursery
IICA	Inter-American Institute for Cooperation in Agriculture
IIEN	IRRI's Industrial Extension Network
IIMI	International Irrigation Management Institute
IITA	International Institute of Tropical Agriculture
ILCA	International Livestock Center for Africa
ILRAD	International Laboratory for Research on Animal Diseases
INCOFORE	International Council for Forestry Research and Extension

INGER	International Network on Genetic Enhancement of Rice
INIAA	Instituto Nacional de Investigaciónes Agropecuaria y Agroforestal
INIBAP	International Network for the Improvement of Banana and Plantain
INRA	Institut National de la Recherche Agronomique
INSFFER	International Network on Soil Fertility and Fertilizer Evaluation for Rice
INSURF	International Network on Soil Fertility and Sustainable Rice Farming
INTSOY	International Soybean Program
IOCCC	International Office of Cocoa, Chocolate and Confectionary Sugar
IPMAT	International Pearl Millet Adaptation Trial
IPMDMN	International Pearl Millet Downy Mildew Nursery
IPMDRTP	International Pearl Millet Disease Resistance Testing Program
IPMRN	International Pearl Millet Rust Nursery
IPMSN	International Pearl Millet Smut Nursery
IPTT	International Progeny Testing Trial
IRBBN	International Rice Bacterial Blight Nursery
IRBN	International Rice Blast Nursery
IRBPHN	International Rice Brown Planthopper Nursery
IRCTN	International Rice Cold Tolerance Nursery
IRDWON	International Rice Deep Water Observational Nursery
IRON	International Rice Observational Nursery
IRRI	International Rice Research Institute
IRRSWON	International Rainfed Rice Shallow Water Observational Nursery
IRRSWYN	International Rainfed Rice Shallow Water Yield Nursery
IRSATON	International Rice Salinity and Alkalinity Tolerance Observational Nursery
IRSBN	International Rice Stemborer Nursery
IRTN	International Rice Tungro Nursery
IRTP	International Rice Testing Program
IRUSS	International Rice Ufra Screening Set
IRWBPHN	International Rice Whitebacked Planthopper Nursery
IRYN	International Rice Yield Nursery
ISNAR	International Service for National Agricultural Research
ISVAT	International Sorghum Variety Adaptation Trial
ISWYN	International Spring Wheat Yield Nursery
ITC	International Trypanotolerance Centre
ITPRON	International Tide-Prone Rice Observational Nursery
IUFRO	International Union of Forestry Research Organizations
IURON	International Upland Rice Observational Nursery
IURYN	International Upland Rice Yield Nursery
IWBP	International Witches' Broom Project
IWSWSN	International Winter × Spring Wheat Screening Nursery
JET	Joint European Torus
MARDI	Malaysian Agricultural Research and Development Institute
MINE	Microbial Information Network for Europe
MIRCENs	Micro-Biological Resources Centers
MOVUSAC	Network on Management of Vertisols under Semi-Arid Conditions in Africa
MSU	Michigan State University
NCSU	North Carolina State University
NIARCs	Networks as international agricultural research centers
NifTAL	Nitrogen Fixation by Tropical Agricultural Legumes
NSF	National Science Foundation
NSSL	National Soil Survey Laboratory
ODA	Overseas Development Administration
ONESASA	Oilcrops Network for Eastern and Southern Africa and South Asia
ORSTOM	Office de la Recherche Scientifique et Technique d'Outre-Mer
OSU	Oregon State University

PANESA	Pasture Network for Eastern and Southern Africa
PCCMCA	Programa Cooperativo Centroamericano para el Mejoramiento de Cultivos Alimenticios
PESTNET	African Regional Pest Management Research and Development Network for Integrated Control of Crop and Livestock Pests
PIN	Pigeonpea International Nurseries
PRACIPA	Programa Regional Andino Cooperativo de Papa
PRAPAC	Programme Regional d'Amélioration de la Culture de Pomme de Terre en Afrique Centrale
PRECODEPA	Programa Regional Cooperative de Papa
PROCIPA	Programa Cooperativo de Investigaciónes en Papa
RDISN	Regional Disease and Insect Screening Nursery
RDTN	Regional Disease Trap Nursery
RESPAO	Réseau d'Études des Systèmes de Production en Afrique de l'Ouest
RISPAL	Rede de Investigación en Sistemas de Producción Animal en Latinoamerica
RNAM	Regional Network for Agricultural Machinery
RRI	Rice Research Institute
RRS	Recherche sur Résistance a Sécheresse
RWYT	Regional Wheat Yield Trial
SADCC	Southern African Development Coordination Committee
SAFGRAD	Semi-Arid Food Grain Research and Development Project
SAPPRAD	Southeast Asian Program for Potato Research and Development
SAREC	Swedish Agency for Research Cooperation with Developing Countries
SCS	Soil Conservation Service
SDC	Swiss Development Cooperation
SDI	Selective Dissemination of Information
SMIC	Sorghum and Millets Information Center
SMSS	Soil Management Support Services
SPAAR	Special Program for African Agricultural Research
SPDC	Special Program for Developing Countries
SUAN	Southeast Asian Universities Agroecosystem Network
TAC	Technical Advisory Committee
UCD	University of California, Davis
UF	University of Florida
UG	University of Georgia
UH	University of Hawaii
UIUC	University of Illinois at Urbana-Champaign
UM	University of Maryland
UN	University of Nebraska
UNDP	United Nations Development Program
UNEP	United Nations Environment Program
UNESCO	United Nations Educational, Scientific and Cultural Organization
USAID	United States Agency for International Development
USDA	United States Department of Agriculture
VEOLA	Latin American Disease and Insect Screening Nursery
WAFSRN	West African Farming Systems Research Network
WARDA	West Africa Rice Development Association
WMO	World Meteorological Association

References

Abelson, P. H., and J. W. Rowe. 1987. "A New Agricultural Frontier." *Science* 235:1450–51.

Asante, S. K. B. 1986. "Food as a Focus of National and Regional Policies in Contemporary Africa." In *Food in Sub-Saharan Africa*, A. Hansen and D. E. McMillan (ed.), pp. 11–24. Lynne Rienner Publishers, Boulder, Colo.

Awakul, S. 1980. "Rice Research and IRTP Involvement in Thailand." In *Rice Improvement in China and Other Asian Countries*, pp. 281–86. International Rice Research Institute/Chinese Academy of Sciences, Los Baños, Philippines.

Baum, W. C. 1986. *Partners against Hunger: The Consultative Group on International Agricultural Research.* World Bank, Washington, D.C.

Beinroth, F. H., G. Uehara, J. A. Silva, R. W. Arnold, and F. B. Cady. 1980. "Agrotechnology Transfer in the Tropics Based on Soil Taxonomy." *Advances in Agronomy* 33:303–39.

Bell, R. D., T. J. Willcocks, D. C. Kemp, and A. A. Metianu. 1985. *Review of the Impact of the Agricultural Engineering Component of the Work of the CGIAR Institutes.* National Institute of Agricultural Engineering, Wrest Park, Silsoe, Bedford, England.

Bockhop, C. W., R. E. Stickney, M. M. Hammond, B. J. Cochran, V. R. Reddy, F. E. Nichols, and S. C. Labro. 1985. "The IRRI Industrial Liaison Program." Paper presented at the International Conference on Agricultural Equipment for Developing Countries, International Rice Research Institute, Los Baños, Philippines, 2–6 September.

Borlaug, N. E. 1983. "Contributions of Conventional Plant Breeding to Food Production." *Science* 219:689–93.

——. 1986. "Accelerating Agricultural Research and Production in the Third World: A Scientist's Viewpoint." *Agriculture and Human Values* 3(3):5–14.

Brady, N. C. 1985. "IRRI in the Next 25 Years: The Future." In *Impact on Science*, pp. 175–82. International Rice Research Institute, Los Baños, Philippines.

Breth, S. A. 1986. *Mainstreams of CIMMYT Research: A Retrospective*. Centro Internacional de Mejoramiento de Maiz y Trigo, El Batán, Mexico.

Brown, L. R., and E. C. Wolf. 1985. *Reversing Africa's Decline*. Worldwatch Institute Paper 65, Washington, D.C.

Brumby, P. J. 1986. *ILCA and Its Strategy*. International Livestock Centre for Africa, Addis Ababa, Ethiopia.

BSP. 1981. *Annual Report, 1980–81*. Benchmark Soils Project, Department of Agronomy and Soil Science, College of Tropical Agriculture and Human Resources, University of Hawaii at Manoa.

——. 1982a. *Procedures and Guidelines for Agrotechnology Transfer Experiments with Maize in a Network of Benchmark Soils*. College of Tropical Agriculture and Human Resources, University of Hawaii, Research Extension Series 15.

——. 1982b. *Assessment of Agrotechnology Transfer in a Network of Tropical Soil Families*. Benchmark Soils Project, Progress Report 2, July 1979–September 1982, Department of Agronomy and Soil Science, College of Tropical Agriculture and Human Resources, University of Hawaii/Department of Agronomy and Soils, College of Agricultural Sciences, University of Puerto Rico.

Bunting, A. H. 1985. "The International Agricultural Research Centers and Agricultural Education in Developing Countries." In: *Education for Agriculture*, pp. 37–50. International Rice Research Institute, Los Baños, Philippines.

Cabanilla, V., and T. Hargrove. 1986. "Copublication in the Third World: Breaking the Language Barrier." *Scholarly Publishing* 17:165–80.

CABI. 1986. "EEC Support for Microbial Culture Databanks." *CABI News* (Commonwealth Agricultural Bureaux International), October, p. 3.

Carangal, V. R. 1988. "International Collaboration in Rice Farming Systems Research." Paper presented at the Food Legume Coordinating Meeting, 30 April–1 May, Bangkok, Thailand.

CGIAR. 1985. *Summary of International Agricultural Research Centers: A Study of Achievements and Potential*. Consultative Group on International Agricultural Research, World Bank, Washington, D.C.

——. 1987. "National Programs: Taking over from the Centers." *News from CGIAR* (Consultative Group on International Agricultural Research, World Bank) 6(3):5.

Chang, T. T., C. R. Adair, and T. H. Johnston. 1982. "The Conservation and Use of Rice Genetic Resources." *Advances in Agronomy* 35:37–91.

CIMMYT. 1979. *International Testing Program in Wheat, Triticale and Barley*. Centro Internacional de Mejoramiento de Maiz y Trigo, El Batán, Mexico.

——. 1981. *CIMMYT Review 1981*. Centro Internacional de Mejoramiento de Maiz y Trigo, El Batán, Mexico.

——. 1985. *CIMMYT Regional Economics Programme: Phase II Baseline Data Summary for USAID REDSO/ESA December 1985 (revised)*. Centro Internacional de Mejoramiento de Maiz y Trigo, Eastern and Southern Africa Region, Nairobi.

——. 1986. *Veery 'S': Bread Wheats for Many Environments*. Centro Internacional de Mejoramiento de Maiz y Trigo, El Batán, Mexico.

CIP. 1984a. *CIP Region VII: Southeast Asia and the Pacific*. Centro Internacional de la Papa, Lima, Peru.

——. 1984b. *Potatoes for the Developing World: A Collaborative Experience*. Centro Internacional de la Papa, Lima, Peru.

——. 1985a. *El Programa Regional Cooperativo de Papa: Informe de la Misión de Revisión, Junio de 1984*. Centro Internacional de la Papa, Lima, Peru.

——. 1985b. "Training Activities." *CIP Circular* (Centro Internacional de la Papa, Lima) 13(4):8–9.

——. 1986. *International Potato Center: Annual Report 1985*. Centro Internacional de la Papa, Lima, Peru.

Collinson, M. 1987. "Farming Systems Research: Procedures for Technology Development." *Experimental Agriculture* 23:365–86.

Coronel, L. V. 1987. "ESA Goes Networking: Region Tests Computer Networking." *Bank's World* (World Bank) 6(3):2–5.

Correa, W. 1986. "Seeds of Promise." *Rotarian*, July, pp. 12–15.

CRIFC. 1986. *Crops-Livestock Systems Research in Upland Agriculture*. Central Research Institute for Food Crops, Bogor, Indonesia.

Crosby, A. W. 1986. *Ecological Imperialism: The Biological Expansion of Europe, 900–1900*. Cambridge University Press, Cambridge.

Cummins, D. G. N.d. *Peanut Collaborative Research Support Program (CRSP): Executive Summary 1985–1989*. U.S. Agency for International Development, Washington, D.C., Grant No. DAN-4048-G-SS-2065-00.

Dalrymple, D. G. 1985. "The Development and Adoption of High-Yielding Varieties of Wheat and Rice in Developing Countries." *American Journal of Agricultural Economics* 67(5):1067–73.

——. 1986. *Development and Spread of High-Yielding Rice Varieties in Developing Countries*. U.S. Agency for International Development, Washington, D.C.

Denning, P. J. 1987. "The Science of Computing: A New Paradigm for Science." *American Scientist* 75:572–73.

Desai, P. N. 1982. "Administration of International Cooperation in Indian Agricultural Research." *Agricultural Administration* 10(1):13–22.

Dickman, S. 1987. "Academic Networks: Taking Stock." *Nature* 328:752–53.

Dickson, D. 1987. "Networking: Better Than Creating New Centers?" *Science* 237:1106–7.

Dubin, H. J., and S. Rajaram, 1982. "The CIMMYT's International Approach to Breeding Disease-Resistant Wheat." *Plant Disease* 66:967–71.

Egan, G. 1988. *Change-Agent Skills A: Assessing and Designing Excellence*. University Associates, San Diego.

Eicher, C. K. 1982. "Facing up to Africa's Food Crisis." *Foreign Affairs* 61(1):151–74.

——. 1988. "Sustainable Institutions for African Agricultural Development." Paper presented at seminar, The Changing Dynamics of Global Agriculture: Research Policy Implications for National Agricultural Research Systems, 22–28 September 1988, Feldafing, Federal Republic of Germany.

Etzioni, A. 1964. *Modern Organizations*. Prentice-Hall, Englewood Cliffs, N.J.

Ezeta, F. N. N.d. "Collaborative Country Research Networks." Centro Internacional de la Papa, Lima.

FAO. 1985. "Cooperative Research Networks in the Near East." Paper presented at the Near East Regional Commission on Agriculture: First Session, 30 March–2 April, Food and Agriculture Organization, Cairo, Egypt.

Foley, D. 1989. "Networking for Social Change: The International Planned Parenthood Federation, 1952–1966." Paper presented at the Annual Conference of the American Society for Public Administration, 8–11 April, Miami, Fla.

France. 1986. *French Priorities in Tropical Food Crop Research*. Ministries of Cooperation and Research/Centre de Cooperation International en Recherche Agronomique pour le Développement (CIRAD)/Institut National de la Recherche Agronomique (INRA)/ORSTOM, Paris.

Francis, C. A. 1986. "Future Perspectives of Multiple Cropping." In *Multiple Cropping Systems*, C. A. Francis (ed.), pp. 351–70. Macmillan, New York.

Fresco, L. O., and S. V. Poats. 1986. "Farming Systems Research and Extension: An Approach to Solving Food Problems in Africa." In *Food in Sub-Saharan Africa*, A. Hansen and D. McMillan (eds.), pp. 305–31, Lynne Rienner Publishers, Boulder, Colo.

FSSP. 1986. "2nd West African Integrated Livestock Systems Networkshop." *Farming Systems Support Project Newsletter* (University of Florida) 4(2):1.

Garwin, L. 1987. "Conflicting Signals from LMC Supernova." *Nature* 326:121.

Greenland, D. J., E. T. Craswell, and M. Dagg. 1987. "International Networks and Their Potential Contribution to Crop and Soil Management Research." *Outlook on Agriculture* 16(1):42–50.

Hailu, M. 1989. "Information Networking in Africa: A Framework for Action." In *Report of the CGIAR Documentation and Information Services Meeting, 16–20 Jan 1989, ICRISAT Center, India*, pp. 14–21. International Crops Research Institute for the Semi-Arid Tropics (ICRISAT), Patancheru, Andhra Pradesh, India.

Hanson, H. 1979. "Plant and Animal Resources for Food Production by Developing Countries in the 1980s." Paper presented at the conference on agricultural production, Bonn, Federal Republic of Germany, October 8–12.

Hardin, L. S., J. Morris, P. Rashid, and S. Ozgediz. 1986. *Report of the First External Management Review of the International Livestock Centre for Africa (ILCA)*. Consultative Group on International Agricultural Research, World Bank, Washington, D.C.

Hayes, H. K. 1963. *A Professor's Story of Hybrid Corn*. Burgess Publishing Co., Minneapolis.

Hepworth, H. M. 1987. "The Human Element: CIMMYT Training." In *CIMMYT's 20th Anniversary: A Commemoration*, pp. 23–24. Centro Internacional de Mejoramiento de Maiz y Trigo, El Batán, Mexico.

Hesterman, B. 1986. "The Dynamics of Pastoral Systems and Ecological Change in Semi-Arid Subsaharan Africa: Implications for Development Policy." M.S. thesis, Department of Geography, University of Florida, Gainesville.

Hogan, E. B., K. O. Rachie, and J. S. Robins. 1986. *Collaborative Research Support: Program Review Study*. Report submitted to the U.S. Agency for International Development, Washington, D.C.

IBSNAT. 1985. *IBSNAT Progress Report 1982–1985*. International Benchmark Sites Network for Agrotechnology Transfer, U.S. Agency for International Development, Washington, D.C./Department of Agronomy and Soil Science, College of Tropical Agriculture and Human Resources, University of Hawaii, Honolulu.

ICRISAT. 1982. *ICRISAT Annual Report 1981*. International Crops Research Institute for the Semi-Arid Tropics, Patancheru, India.

——. 1984. *The Seventh International Pearl Millet Adaptation Trial (IPMAT 7), 1981*. International Crops Research Institute for the Semi-Arid Tropics, Patancheru, India.

——. 1985. *ICRISAT Research Highlights 1984*. International Crops Research Institute for the Semi-Arid Tropics, Patancheru, India.

——. 1986. *ICRISAT Research Highlights 1985*. International Crops Research Institute for the Semi-Arid Tropics (ICRISAT), Patancheru, Andhra Pradesh, India.

——. N.d. *International Pearl Millet Disease Resistance Testing Program (IPMDRTP): Report of the Eighth (1983) International Pearl Millet Downy Mildew Nursery (IPMDMN)*. International Crops Research Institute for the Semi-Arid Tropics, Patancheru, India.

IDRC. 1986. *Searching: IDRC 1985—Research: A Path to Development*. International Development Research Center, Ottawa.

IFDC. 1985. *The IFDC Story*. International Fertilizer Development Center, Muscle Shoals, Ala.

ILCA. 1981. *ILCA's Longer Term Plan*. International Livestock Center for Africa, Addis Ababa, Ethiopia.

——. 1986a. *ILCA Annual Report 1985/6*. International Livestock Center for Africa, Addis Ababa, Ethiopia.

——. 1986b. *The ILCA/ILRAD Trypanotolerance Network: Situation Report, December 1985, Proceedings of a Network Meeting Held at ILCA, Nairobi*. International Livestock Center for Africa, Addis Ababa, Ethiopia.

——. 1988. "ILCA and NARS Partners—Responses to the Biennial Meeting." *ILCA Newsletter* (International Livestock Center for Africa) 7(2):2–3.

ILCA/FAO/UNEP. 1979. *Trypanotolerant Livestock in West and Central Africa*. International Livestock Center for Africa, Addis Ababa, ILCA Monograph No. 2, 2 vols.

ILRAD. 1986a. "Trypanotolerance Network Meeting." *ILRAD Reports* (International Laboratory for Research on Animal Diseases) 4(1):5–6.

——. 1986b. "Trypanosomiasis Research." *ILRAD Reports* (International Laboratory for Research on Animal Diseases) 4(3):2–3.

——. 1986c. *ILRAD: 1985*. International Laboratory for Research on Animal Diseases, Nairobi, Kenya.

——. 1987. "Improved Trypanosomiasis Control: Studies on Drug Treatment." *ILRAD Reports* (International Laboratory for Research on Animal Diseases) 5(1):1–6.

——. 1988. "Trypanosomiasis." *ILRAD Highlights 1988* (International Laboratory for Research on Animal Diseases), pp. 2–4.

INIBAP. 1986. *Banana Research in Eastern Africa: Proposal for a Regional*

Research and Development Network. International Network for the Improvement of Banana and Plantain (INIBAP), Montpellier, France.

——. 1987. *Plantain in West and Central Africa: Proposal for a Regional Research and Development Network.* International Network for the Improvement of Banana and Plantain (INIBAP), Montpellier, France.

INSFFER. 1986. "Proceedings of the ICAR-IRRI Planning Meeting on INSFFER Held at Narendra University of Agriculture and Technology, Kumargunj, U.P., India, on April 18 1986." International Network on Soil Fertility and Fertilizer Evaluation for Rice, International Rice Research Institute, Los Baños, Philippines.

IRRI. 1984. *Multilanguage Publication in Agriculture: Workshop Report and Description of Participating Agencies.* International Rice Research Institute/International Development Research Centre, Los Baños, Philippines.

——. 1985a. *IRRI Highlights 1984.* International Rice Research Institute, Los Baños, Philippines.

——. 1985b. *1985 Annual Report: The International Rice Testing Program.* International Rice Research Institute, Los Baños, Philippines.

——. 1985c. *Farming Systems Research at IRRI: An Overview.* International Rice Research Institute, Los Baños, Philippines.

——. 1986. *IRRI Highlights 1985: Accomplishments and Challenges.* International Rice Research Institute, Los Baños, Philippines.

——. 1988. *IRRI Strategy, 1990–2000.* International Rice Research Institute, Los Baños, Philippines.

ISNAR. 1985. *Regional Research Networks: The Experience of PRECODEPA.* International Service for National Agricultural Research, The Hague.

Jaramillo, R. (ed.). 1988. *Investigación en banana y platano en America Latina y el Caribe: Propuesta para la Organización y Desarollo de la Red Regional.* International Network for the Improvement of Banana and Plantain (INIBAP), San José, Costa Rica.

Jayaraman, K. S. 1987. "India Establishes Countrywide Satellite Computerized Database." *Nature* 325:753.

Jennings, D. 1987. "Computing the Best for Europe." *Nature* 329:775–78.

——, L. H. Landweber, I. H. Fuchs, D. J. Farber, and W. R. Adrion. 1986. "Computer Networking for Scientists." *Science* 231:943–50.

Jin-Hua, S. 1980. "Rice Breeding in China." In *Rice Improvement in China and Other Asian Countries*, pp. 9–30. International Rice Research Institute/Chinese Academy of Agricultural Sciences, Los Baños, Philippines.

Kauffman, H. E., M. J. Rosero, and V. R. Carangal. 1982. "International Networks." In *Rice Research Strategies for the Future*, pp. 503–25. International Rice Research Institute, Los Baños, Philippines.

Khush, G. S. 1984. "IRRI Breeding Program and Its Worldwide Impact on Increasing Rice Production." In *Gene Manipulation in Plant Improvement*, J. P. Gustafson (ed.), pp. 61–94. Plenum Press, New York.

Killman, R., and M.J. Saxton. 1984. *Beyond the Quick Fix.* Jossey-Bass, San Francisco.

Kolata, G. 1986. "Mystery Disease at Lake Tahoe Challenges Virologists and Clinicians." *Science* 234:541–42.

Kotter, J. P. 1988. *The Leadership Factor*. Free Press, New York.

Kristofferson, D. 1987. "The BIONET Electronic Network." *Nature* 325:555–56.

Lamung, C. B. 1985. "Rapid Community Appraisal of Upland Agroecosystems." In *Agroecosystem Research in Rural Resource Management and Development*, P. E. Sajise, A. T. Rambo, C. M. Rebancos (eds.), pp. 94–104. Program on Environmental Science and Management, University of the Philippines at Los Baños and the Southeast Asian Universities Agroecosystem Network.

Lipnack, J., and J. Stamps. 1987. *The Networking Book: People Connecting with People*. Routledge & Kegan Paul, New York.

Litwak, E., and L. F. Hylton. 1962. "Interorganizational Analysis: A Hypothesis on Co-ordinating Agencies." *Administrative Science Quarterly* 6:395–420.

Loegering, W. Q., and N. E. Borlaug. 1966. *Contribution of the International Spring Wheat Rust Nursery to Human Progress and International Good Will*. Agricultural Research Service, USDA, Washington, D.C.

Maddox, J. 1987. "New European Collaborators." *Nature* 330:417.

Mamaril, C. P. 1984. "The INSFFER Program: Its Role in Rice Production." In *Proceedings of the Fifth ASEAN Soil Conference, Bangkok, Thailand, 10–23 June 1984*, S. Panichapong, C. Niamskul, A. Promprasit, and M. Newport (eds.), vol. J3, pp. 1–15. Mapping and Printing Division, Department of Land Development, Ministry of Agriculture and Co-operatives, Bangkok, Thailand.

——. 1985. "International and National Cooperation in Long-Term Coordinated Schemes of Experimentation on Fertilizers." In *Potassium in the Agricultural Systems of the Humid Tropics*, pp. 287–95. Nineteenth Colloquium of the International Potash Institute, 25–29 November 1985, Bangkok, Thailand.

Mashler, W. T. 1985. "The Role of the International Agricultural Research Centers in Cooperative Research." In *Impact of Science on Rice*, pp. 241–45. International Rice Research Institute, Los Baños, Philippines.

Moseman, A. H. 1970. *Building Agricultural Research Systems in the Developing Nations*. Agricultural Development Council, New York.

Murray, M., D. J. Clifford, G. Gettinby, W. F. Snow, and W. I. M. McIntyre. 1981. "Susceptibility to African Trypanosomiasis of N'Dama and Zebu Cattle in an Area of *Glossina moristans submoristans* Challenge." *Veterinary Record* 109:503–10.

——, W. I. Morrison, and D. D. Whitelaw. 1982. "Host Susceptibility to African Trypanosomiasis: Trypanotolerance." In *Advances in Parasitology*, J. R. Baker and R. Muller (eds.), 1:pp. 1–68. Academic Press, London.

——, J. C. M. Trail, D. A. Turner, and Y. Wissocq. 1983. *Livestock Productivity and Trypanotolerance: Network Training Manual*. International Livestock Center for Africa, Addis Ababa, Ethiopia.

Oram, P. 1980. *Collaboration between National and International Institutions in the Development of Improved Agricultural Technology for Low-Income Countries*. International Food Policy Research Institute, Washington, D.C.

——. 1988. *International Agricultural Research Needs in Sub-Saharan Africa: Current Problems and Future Imperatives—Issues and Options for the CGIAR*. International Food Policy Research Institute, Washington, D.C.

Osmond, B. 1986. "Research in South-East Asia." *Nature* 320:307–8.

Oyekan, J. 1987. "IITA Marks 20th Anniversary." *IITA Briefs* (International Institute of Tropical Agriculture) 8(3):1–2.

Ozgediz, S. 1987. "Lessons on Board Performance." Memorandum, Consultative Group on International Agricultural Research, World Bank, Washington, D.C., 25 March.

——. 1988. "Point of Views, Concepts and Issues in Strategic Planning." *CIMMYT 1987 Annual Report* (International Maize and Wheat Improvement Center), pp. 8–13.

——. 1990."Overview of Management in CGIAR Centers." Paper presented at the Mid-Year Meeting of the Consultative Group on International Agricultural Research (CGIAR), 21–25 May, The Hague, Netherlands.

Pande, H. K., and R. Seetharam. 1980. "Rice Research and Testing Program in India." In *Rice Improvement in China and Other Asian Countries*, pp. 37–49. International Rice Research Institute/Chinese Academy of Agricultural Sciences, Los Baños, Philippines.

Pascale, R. T., and A. G. Athos. 1981. *The Art of Japanese Management*. Warner Books, New York.

Pathak, M. D. 1985. "Training Implications of Recent Progress in Rice Research." In *Education for Agriculture*, pp. 51–71. International Rice Research Institute, Los Baños, Philippines.

Paul, S. 1982. *Managing Development Programs*. Westview Press, Boulder, Colo.

Plucknett, D. L., and N. J. H. Smith. 1982. "Agricultural Research and Third World Food Production." *Science* 217:215–20.

——. 1984. "Networking in International Agricultural Research." *Science* 225:989–93.

——. 1986a. "International Cooperation in Cereal Research." In *Advances in Cereal Science and Technology*, vol. 8, Y. Pomeranz (ed.), pp. 1–14. American Association of Cereal Chemists, St. Paul, Minn.

——. 1986b. "Sustaining Agricultural Yields: As Productivity Rises, Maintenance Research Is Needed to Uphold the Gains." *Bioscience* 36:40–45.

——. 1987. "Networking as a Research Facilitator." In *ICLARM Report 1986*, J. L. Maclean and L. B. Dizon (eds.), pp. 25–31. International Center for Living Aquatic Resources Management, Manila, Philippines.

Plucknett, D. L., N. J. H. Smith, J. T. Williams, and N. M. Anishetty. 1987. *Gene Banks and the World's Food*. Princeton University Press, Princeton.

Rambo, A. T., and P. E. Sajise. 1985. "Developing a Regional Network for Interdisciplinary Research on Rural Ecology: The Southeast Asian Universities Agroecosystem Network (SUAN) Experience." *Environmental Professional* 7:289–98.

Reddy, V. R. 1984. "IRRI Industrial Extension Project in Indonesia from March 1978 to August 1984." Paper presented at seminar on industrial extension sponsored by the Chinese Academy of Agricultural Mechanization Sciences, Beijing, Peoples' Republic of China, October.

Rijk, A. G. 1985. "Credit for Agricultural Mechanization in Asia." *RNAM [Regional Network for Agricultural Machinery] Newsletter* 22:19.

RNAM. 1985. *India, Indonesia, Islamic Republic of Iran, Pakistan, Philippines,*

Republic of Korea, Sri Lanka, Thailand. Regional Network for Agricultural Machinery, Los Baños, Philippines.

——. 1986. "Strengthening Institutional Capabilities: RNAM Organizes Training, Workshops, Technical Meetings and Study Tours." *RNAM [Regional Network for Agricultural Machinery] Newsletter* 25:4–5.

Rodrigues, C. J. 1977. "Coffee (*Coffea* spp.)." In *Plant Health and Quarantine in International Transfer of Genetic Resources*, W. B. Hewitt and L. Chiarappa (eds.), pp. 137–54. CRC Press, Cleveland.

Roger, P. A., J. K. Ladha, and I. Watanabe. 1985. "Cooperative Aspects of the Research Programs on Biological Nitrogen Fixation at IRRI." Paper presented at the International Rice Research Conference, Los Baños, Philippines, 1–5 June.

RRI. N.d. *The RD Rice Varieties in Thailand*. Rice Research Institute, Department of Agriculture, Ministry of Agriculture and Cooperatives, Bangkok, Thailand.

Sabin, A. B. 1986. *Role of My Cooperation with Soviet Scientists in the Conquest of Polio: Some Lessons and Challenges*. Twenty-Third Cosmos Club Award Lecture, Cosmos Club, Washington, D.C.

Sarma, J. S. 1986. *Cereal Feed Use in the Third World: Past Trends and Projections to 2000*. International Food Policy Research Institute, Washington, D.C., Research Report 57.

Satari, G. 1986. "Message of the Director General of AARD at the INSFFER Planning Meeting Sukamandi, 21–22 January 1986." In *Report on the INSFFER Collaborators' Meeting in Indonesia, January 21–22, 1986, Sukamandi, Indonesia*, C. P. Mamaril (ed.), International Rice Research Institute, Los Baños, Philippines.

Seshu, D. V. 1986. *An Overview of the International Rice Testing Program in Africa, 1975–85*. International Rice Research Institute, Los Baños, Philippines.

——. 1988. "Agricultural Research Networks—A Model for Success." In *Vegetable Research in Southeast Asia*, pp. 211–18. Asian Vegetable Research and Development Center, Shanhua, Taiwan,

——. N.d. *International Rice Testing Program—A Mechanism for International Cooperation in Rice Improvement Coordinated by the International Rice Research Institute*. International Rice Research Institute, Los Baños, Philippines.

Seshu, D. V., and H. E. Kauffman. 1980. "Differential Response of Rice Varieties to the Brown Planthopper in International Screening Tests." *IRRI (International Rice Research Institute) Research Paper Series* 52.

Simpson, J. R., and R. E. McDowell. 1986. "Livestock in the Economies of Sub-Saharan Africa." In *Food in Sub-Saharan Africa*, A. Hansen and D. McMillan (eds.), pp. 207–21. Lynne Rienner Publishers, Boulder, Colo.

Singh, B. 1985. "Rice Production in India." In *Impact of Science on Rice*, pp. 75–80. International Rice Research Institute, Los Baños, Philippines.

Siwi, B. H., and H. M. Beachell. 1980. "The GEU Concept in Indonesia." In *Rice Improvement in China and Other Asian Countries*. International Rice Research Institute/Chinese Academy of Agricultural Sciences, Los Baños, Philippines.

Smith, N. J. H. 1986. *Botanic Gardens and Germplasm Conservation*. Harold L. Lyon Arboretum Lecture Series No. 14. University of Hawaii Press, Honolulu.

SPAAR. 1986. *African Agricultural Research Networks: Summary Papers and Tables.* Meeting of the Technical Working Group on Networking of the Special Program for African Agricultural Research (SPAAR), Brussels, Belgium, July 7–8.

Spears, J. 1985. "Bank Support for Research in Africa." Memorandum to E. Stern, Vice-President, Operations Policy, World Bank, Washington, D.C.

Sprague, E. W., and R. L. Paliwal. 1984. "CIMMYT's Maize Improvement Programme." *Outlook on Agriculture* 13(1):24–31.

Swanson, B. E. 1975. *Organizing Agricultural Technology Transfer: The Effects of Alternative Arrangements.* Program of Advanced Studies in Institution Building and Technical Assistance Methodology, Indiana University, Bloomington.

TAC. 1986a. *TAC Review of CGIAR Priorities and Future Strategies: Notes on the Issues Raised during the Discussion at ICW, Washington, 31 October 1985 (Agenda Item 9b).* Technical Advisory Committee Secretariat, Consultative Group on International Agricultural Research, Rome.

——. 1986b. *Training in the CGIAR System: Building Human Resources for Research to Improve Food Production in Developing Countries.* Technical Advisory Committee Secretariat, Consultative Group on International Agricultural Research, Rome.

Tichy, N. M. 1983. *Managing Strategic Change.* Wiley, New York.

Timothy, D. H., P. H. Harvey, and C. R. Dowswell. 1988. *Development and Spread of Improved Maize Varieties and Hybrids in Developing Countries.* U.S. Agency for International Development, Washington, D.C.

Valverde, C. 1988. *Agricultural Research Networking: Development and Evaluation.* International Service for National Agricultural Research, The Hague.

Van de Ven, A. H., and D. L. Ferry. 1980. *Measuring and Assessing Organizations.* Wiley, New York.

Vasal, S. K., A. Ortega, and S. Pandey. 1982. *CIMMYT's Maize Germplasm Management, Improvement, and Utilization Program.* Centro Internacional de Mejoramiento de Maiz y Trigo, El Batán, Mexico.

Vaz, W. L. C. 1987. "Problems of Portuguese Science." *Nature* 326:238.

Vermeer, D. E. 1986. "Dwarf Cattle and Trypanosomiasis." *Science* 218: 636.

Walsh, J. 1986a. "River Blindness: A Gamble Pays Off." *Science* 232:922–25.

——. 1986b. "Crop Research Network Makes Some Changes." *Science* 234:1190–91.

Wennergren, E. B., D. L. Plucknett, N. J. H. Smith, W. L. Furlong, and J. H. Joshi. 1986. *Solving World Hunger: The U.S. Stake.* Consortium for International Cooperation in Higher Education, Washington, D.C./Seven Locks Press, Cabin John, Md.

Whitmore, T. C. 1985. *Tropical Rain Forests of the Far East.* Clarendon Press, Oxford.

Winkelmann, D. L. 1987. "Networking: Some Impressions from CIMMYT." In *The Impact of Research on National Agricultural Development: First International Meeting of National Agricultural Research Systems and the Second IFARD Global Convention*, B. Webster and C. Valverde (eds.), pp. 131–45. International Service for National Agricultural Research, The Hague.

Wolf, E. C. 1986. *Beyond the Green Revolution: New Approaches for Third World Agriculture*. Worldwatch Institute Paper 73, Washington, D.C.

Wright, K. 1986. "Insect Virus as Super-Vector?" *Nature* 321:718.

Yudelman, M. 1985. *The World Bank and Agricultural Development—An Insider's View*. World Resources Institute, Washington, D.C.

Zaman, S. M. H. 1980. "The Importance of the IRTP in BRRI's Varietal Improvement Work." In *Rice Improvement in China and Other Asian Countries*, pp. 251–59. International Rice Research Institute/Chinese Academy of Agricultural Sciences, Los Baños, Philippines.

Zandstra, H. G. 1986. "Canadian Support to Agricultural Research for the Developing World." Paper presented at the CGIAR mid-term meeting, 19–23 May, Ottawa, Canada.

Author Index

General Index